AAPS Advances in the Pharmaceutical Sciences Series

The AAPS Advances in the Pharmaceutical Sciences Series, published in partnership with the American Association of Pharmaceutical Scientists, is designed to deliver volumes authored by opinion leaders and authorities from around the globe, addressing innovations in drug research and development, and best practice for scientists and industry professionals in the pharma and biotech industries.

Volume 31

Series Editor

Yvonne Perrie, Strathclyde Institute of Pharmacy, University of Strathclyde, Bearsden, Dunbartonshire, UK

More information about this series at http://www.springer.com/series/8825

Abdul W. Basit • Simon Gaisford
Editors

3D Printing of Pharmaceuticals

aapspress

Springer

Editors
Abdul W. Basit
Department of Pharmaceutics
UCL School of Pharmacy
University College London
London, UK

FabRx Ltd.
Ashford, Kent, UK

Simon Gaisford
Department of Pharmaceutics
UCL School of Pharmacy
University College London
London, UK

FabRx Ltd.
Ashford, Kent, UK

ISSN 2210-7371 ISSN 2210-738X (electronic)
AAPS Advances in the Pharmaceutical Sciences Series
ISBN 978-3-030-08091-4 ISBN 978-3-319-90755-0 (eBook)
https://doi.org/10.1007/978-3-319-90755-0

Printed on acid-free paper

This Springer imprint is published by the registered company Springer International Publishing AG part of Springer Nature.
The registered company address is: Gewerbestrasse 11, 6330 Cham, Switzerland

Preface

The world has experienced three major industrial revolutions with each one changing the way that we live, work and relate to one another. The first began in the late eighteenth century with the advent of steam engines, the textile industry and mechanised factories. The second transition involved the mass production of goods powered by electricity whilst the third industrial revolution was established by the adoption of automation. We now stand on the brink of a fourth industrial revolution where technologies such as cloud-based computing, contactless payment, artificial intelligence and three-dimensional (3D) printing, which were once a fantasy, have become a reality. The recent development of 3D printing and additive manufacturing technologies has led to a seemingly boundless number of applications in many fields, ranging from engineering and architecture through to fashion and medicine.

More recently, 3D printing has expanded into the pharmaceutical sector which has been forecasted to cause a paradigm shift in the way that we design, manufacture and use medicines. Conventional drug delivery systems were first introduced around 200 years ago and have generally remained the same in the form of tablets and capsules. With the emergence of 3D printing has come the idea of printable tablets – Printlets® – enabling the production of a wide range of dosage form designs and geometries. The first chapter of this book reviews the revolutionary potential of 3D printing within the pharmaceutical industry and clinical pharmacy practice.

Chapter 2 traces the history, evolution of concepts and achievements that has brought the discipline to where it is today. Binder jet printing, unlike other 3D printing technologies that use liquids, gels or pastes, fabricates objects by fusing powdered materials in a layer-by-layer approach. The first 3D printed pharmaceutical product approved by the U.S. Food and Drug Administration (FDA), Spritam®, was achieved using this method and has introduced the feasibility of transitioning 3D printed techniques for scale-up manufacturing. Chapter 3 provides such perspectives of this pharmaceutical 3D printing endeavour.

In 1984 the first additive manufacturing technology, known as stereolithography, was invented by Charles Hull, who is recognised as the father of 3D printing. Applications of this technique are discussed from an industrial perspective highlighting pharmaceutical opportunities in Chapter 4, ranging from the 3D printing of noses to intricate Printlet® structures.

Indeed, the 1980s was somewhat a decade for technological advancement for the scope of 3D printing. Complexity was something to be embraced and no longer avoided due to the invention of the selective laser sintering technique. Developed by Carl Deckard and Joe Beaman, this technology led to the development of selective laser melting and direct metal laser sintering. Chapter 5 is dedicated to the history of the technology, its current applications in industry and its potential for medical innovation such as tissue engineering.

Due to its ability to incorporate a wide range of materials, the fused deposition modeling technique is considered to be the most applicable 3D printing technology. Its uses are broad ranging across the educational, mechanical, automotive, aerospace, maritime, entertainment, architectural and pharmaceutical industries. Designers from all fields are able to envision, outline and rapidly fabricate their imaginations to life. Chapter 6 provides such perspectives that explore the most widely used 3D printing technique available.

From the open-source community that surround the 3D printing industry, semisolid extrusion printing has been adapted for the fabrication of confectionary to even living cells. The unique attributes of semisolid extrusion printing are beginning to be applied to drug development with increasing possibilities to create on-demand personalised dosages in a clinical setting. Chapter 7 outlines the untapped potential for pharmaceutics.

A chapter devoted to a new arena of additive manufacturing techniques is further discussed in this book, termed as four-dimensional (4D) printing. Using time as the fourth-dimension, 4D printing can create products that independently alter their morphology in response to certain external stimuli. A significant and exciting application of 4D printing is that of bioprinting, reviewed in Chapter 8.

Pioneering clinical feasibility perspectives are provided in Chapter 9 where current 3D printing applications in the medical field are acknowledged following the fabrication of aesthetically authentic prosthetics and even the printing of organs fit for human transplantation.

With an endless portfolio of current and potential applications, it is essential to consider the aspects and issues of intellectual property that surround the 3D printing space. These characteristics are extensively deliberated in Chapter 10.

With the advent of any other avant-garde technology, and as is especially the case in medicine and pharmaceuticals, regulatory requirements also apply to the constantly evolving field of 3D printing, rendering the development of new applications within both an ethically and safety-specific grey area. A concluding chapter (Chapter 11) from the perspective of the FDA is dedicated to the regulatory management surrounding 3D printing.

The 3D printing revolution, like any other, is the product of vision and implementation, brought upon those who were brave enough to make it happen. With the acknowledgement, increasing curiosity and understanding of the capabilities of 3D printing, we stand on the brink of a revolution in digital fabrication with the potential to disrupt how future pharmaceutical products are made.

London, UK

Abdul W. Basit
Simon Gaisford

Contents

Contributors

Atheer Awad Department of Pharmaceutics, UCL School of Pharmacy, University College London, London, UK

Abdul W. Basit Department of Pharmaceutics, UCL School of Pharmacy, University College London, London, UK

FabRx Ltd., Ashford, Kent, UK

Fabrizio Fina Department of Pharmaceutics, UCL School of Pharmacy, University College London, London, UK

Jack Firth Department of Pharmaceutics, UCL School of Pharmacy, University College London, London, UK

Intract Pharma Ltd., London, UK

Simon Gaisford Department of Pharmaceutics, UCL School of Pharmacy, University College London, London, UK

FabRx Ltd., Ashford, Kent, UK

Alvaro Goyanes FabRx Ltd., Ashford, Kent, UK

Grace B. Hatton Department of Pharmaceutics, UCL School of Pharmacy, University College London, London, UK

Catherine Jewell Beck Greener, London, UK

Akm Khairuzzaman Office of Pharmaceutical Quality, The US FDA, Silver Spring, MD, USA

Christine M. Madla Department of Pharmaceutics, UCL School of Pharmacy, University College London, London, UK

Pamela Robles Martinez Department of Pharmaceutics, UCL School of Pharmacy, University College London, London, UK

James Stones Beck Greener, London, UK

Sarah J. Trenfield Department of Pharmaceutics, UCL School of Pharmacy, University College London, London, UK

About the Editors

Abdul W. Basit is Professor of Pharmaceutics at the UCL School of Pharmacy, University College London. Abdul's research sits at the interface between pharmaceutical science and gastroenterology, forging links between basic science and clinical outcomes. His research has been translated into the design of new technologies and improved disease treatments, many of which have been commercialised. Abdul is also a serial entrepreneur and has filed multiple patents, is the recipient of several research awards and has founded three companies (Kuecept, Intract Pharma and FabRx). He is a frequent speaker at international conferences, serves as a consultant to many pharmaceutical corporations and is on the advisory boards of scientific journals, healthcare organisations and charitable bodies.

Simon Gaisford is Professor of Pharmaceutics at the UCL School of Pharmacy, University College London. His research interests focus on the design, manufacture and characterisation of new pharmaceutical products, and he has pioneered pharmaceutical applications of many techniques, including thermal methods, ink-jet printing and additive manufacturing. He has published 110 papers, 8 book chapters and 4 authored books, is the recipient of multiple research awards and has translated his work to the commercial sector founding two companies (Kuecept and FabRx).

Chapter 1
The Shape of Things to Come: Emerging Applications of 3D Printing in Healthcare

Sarah J. Trenfield, Christine M. Madla, Abdul W. Basit, and Simon Gaisford

Abstract We now stand on the brink of a fourth industrial revolution. By the remarkable technological advancements of the twenty-first century, manufacturing is now becoming digitalised. In the last decade, the rise of rapid prototyping has provided individual patient care, acted as an educational and training tool and contributed to research. Innovative technologies such as three-dimensional printing (3DP), have the potential to cause a paradigm shift in medicine design, manufacture and use. Instead of using conventional large batch processes, customised printlets (3D printed tablets) with a tailored dose, shape, size and release characteristics could be produced on-demand. Arguably, never before has the pharmaceutical industry experienced such a transformative technology in medicines manufacture. Indeed, this technology could be utilised throughout the drug development process, ranging from pre-clinical development and first-in-human clinical trials through to front-line medical care (personalized medicines). This chapter aims to discuss the current and future potential applications of 3DP in healthcare and, ultimately, the power of 3DP in pharmaceuticals.

Keywords 3D printing · Additive manufacturing · Pharmaceutics · Bioprinting · Drug delivery systems · Digital health

S. J. Trenfield (✉) · C. M. Madla
Department of Pharmaceutics, UCL School of Pharmacy, University College London, London, UK
e-mail: sarah.trenfield.16@ucl.ac.uk

A. W. Basit · S. Gaisford
Department of Pharmaceutics, UCL School of Pharmacy, University College London, London, UK

FabRx Ltd., Ashford, Kent, UK

A. W. Basit, S. Gaisford (eds.), *3D Printing of Pharmaceuticals*,
AAPS Advances in the Pharmaceutical Sciences Series 31,
https://doi.org/10.1007/978-3-319-90755-0_1

1 3D Printing: The Next Industrial Revolution

Throughout history, the world has undergone three major industrial revolutions relating to manufacturing and production. The first transition began in the late eighteenth century where water and steam was utilised to mechanise production [1]. The second industrial revolution occurred in the early twentieth century by harnessing electrical power to permit mass production. Subsequently, the third industrial revolution was established by the adoption of automation. We now stand on the brink of a fourth industrial revolution. Through the remarkable technological advancements of the twenty-first century, manufacturing is now becoming digitalised. Three-dimensional (3D) printing, an additive manufacturing technique, is set to become a major disruptive technology by enabling the production of bespoke objects of virtually any shape and size, layer by layer. In particular, structures can be created from a digital 3D file using computer-aided design (CAD) software or imaging techniques, such as magnetic resonance imaging (MRI) to readily manufacture individualised objects.

Since the introduction of 3D printing nearly three decades ago, this technology has transformed manufacturing in a boundless field of applications. To this day, 3D printing is often employed to create engineering prototypes due to its fast production speed and cost-effectiveness and was in fact first invented for this purpose. Indeed, it seems that the applications of 3D printing is limited only by the imagination with reports of car parts, customised fashion accessories, organs and even houses being produced using this technology [2].

The applications of 3D printing do not stop there. Indeed, 3D printing is set to become a revolutionary technology within healthcare; due to its capability to produce bespoke and individualised objects, personalised medical prosthetics and devices can be tailored to the individual needs of each patient [3]. Various constructs have already been prepared using 3D printing in the medical field ranging from implants, cardiovascular stents, blood vessels [4], bone repair [5], cells and even 3D printed drug products (see Chap. 3) [6–9].

1.1 3D Printing in Medicine: An Overview

Although this book aims to concentrate on the emerging applications and opportunities of 3D printing in pharmaceutical manufacturing, it is important to highlight how 3D printing has already changed the face of healthcare. In the last decade, the rise of rapid prototyping has provided individualised patient care, contributed to research and acted as an educational and training tool.

3D printing has been introduced in the surgical space as a tool to aid the understanding of complex underlying anomalies and thus, improve diagnostic quality and contribute to pre-surgical planning. 3D printing has already lent itself to simulate craniofacial, maxillofacial [10], spinal [11] and cardiovascular surgeries [12] which provided significant improvements in diagnosis and reduced intraoperative complications due to advanced prototyping techniques.

3D printing has also been applied for the reconstruction of various anatomical structures such as customised limb or facial prostheses, dental or bone restorations [13]. Figure 1.1 outlines the current applications of 3D printing in medicine, however, a more detailed discussion of the materials used, FDA approved products and emerging opportunities are available in Chap. 9.

2 Personalised Medicines

3D printing is also likely to create a paradigm shift in medicines manufacture. Pharmaceutical manufacturing processes were first introduced over 200 years ago although many of which are still in use today (e.g. direct compaction and encapsulation). Whilst these methods are cost-effective for large-scale manufacture, they can be inherently time consuming, labour intensive and dose inflexible. Using novel innovative technologies such as 3D printing could revolutionise the way we make tablets (herein known as printlets: a term that we have coined to describe 3DP oral dosage forms) and thus, move away from a 'one-size-fits-all' approach to permit the personalisation of medicines.

Due to the potential for fabrication of tailored and precise medicines, this technology could be utilised for dispensing on the front-line, as well as throughout the drug development timeline, ranging from pre-clinical development to first-in-human (FIH) clinical trials [17]. In both cases, printlets could be produced to suit the trial or patient requirements (e.g. shape, size, formulation type and dosage). Arguably, never before has the pharmaceutical industry experienced such a transformative technology in medicines manufacture. As such, this chapter will aim to discuss the major pharmaceutical applications of 3D printing, with a focus on the ability of 3D printing to change the way we design, use and manufacture medicines in the clinical and drug development arena [18]. The different 3D printing technologies themselves will be explored and covered in subsequent chapters to come.

Since the implementation of the Precision Medicines Initiative in the U.S. in 2015, there has been increasing emphasis placed on moving medical treatment away from a 'one-size-fits-all' approach towards personalisation [19]. Conventionally, tablets are mass manufactured in a few discrete strengths which are selected from early phase clinical trials and represent the dose required for a safe and therapeutic effect in the majority of the population [20]. However, it is evident that one dose might not fit all; dose requirements can vary based on a patient's genetic profile, disease state and other factors (e.g. gender [21], age [22] and weight) [23–26]. This understanding fashioned the field of personalised medicine, commonly involving the tailoring of therapies to a patient based on their individual characteristics, needs and preferences during all phases of treatment (including prevention, diagnosis, treatment and follow-up) [27].

3DP could help to drive this field of personalised medicine due to its capabilities of producing individualised printlets in small or 'one-off' batches, directly at the

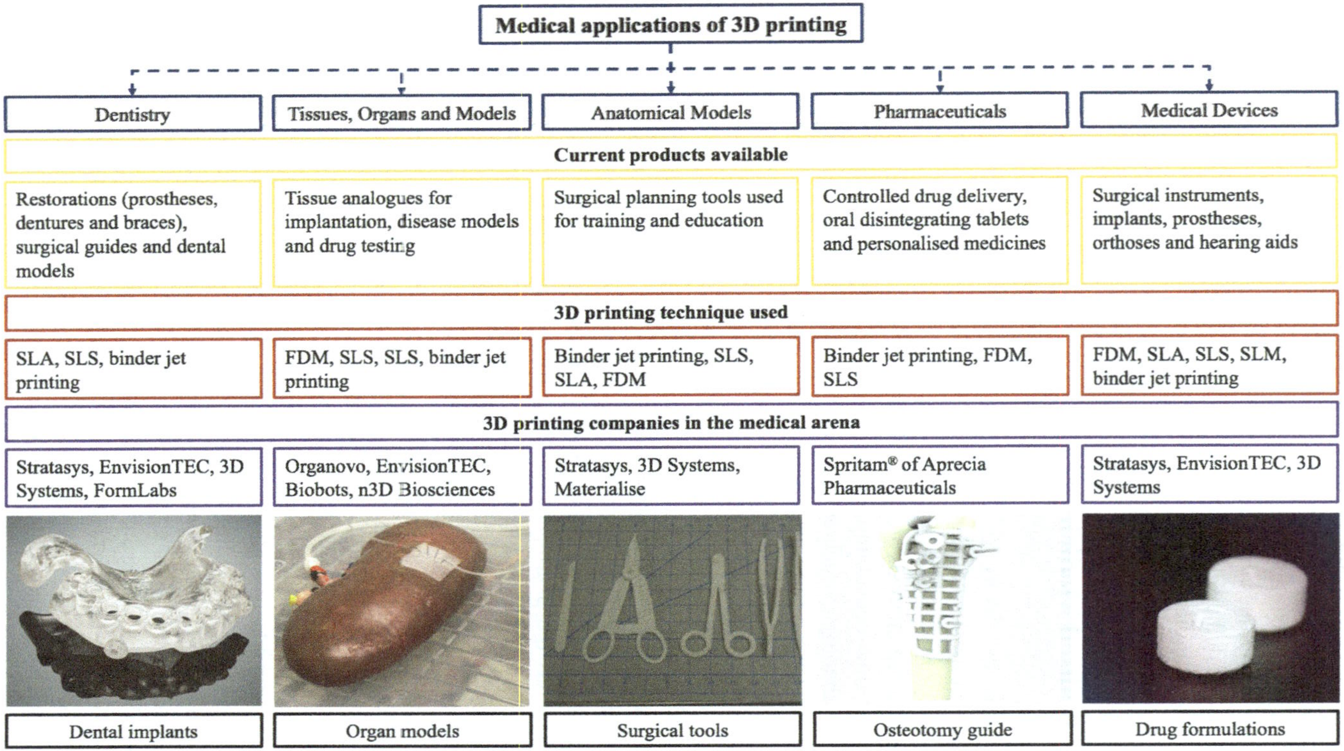

Fig. 1.1 Applications of 3D printing in medicine, examples of products available, method of 3D printing manufacture and companies in the field. (Images reprinted with permissions from [7, 14–16])

point of care. The dosage form could be designed to contain an appropriate dosage, drug combinations, formulation type and/or aesthetics that are tailored to suit the patient [28]. This process could benefit a number of different situations and patient groups on the front-line, which will be discussed in turn.

2.1 *Dose Personalisation*

In young children and the elderly, dosing requirements can be markedly different when compared with adults due to differences in physical characteristics (e.g. body weight, surface area and age) and pharmacokinetics (e.g. metabolic capacity, drug clearance and organ function) [29, 30]. Furthermore, certain medicines with complex dosage regimes, such as narrow therapeutic index drugs [20], require exact dosing to maintain treatment efficacy and patient safety. This can be particularly problematic for those medicines that are formulated as single discrete strengths or formulation types. As such, it has become common practice for patients or carers to split tablets or for staff to extemporaneously prepare formulations (e.g. by crushing licensed tablets or by opening capsules and using the contents) to achieve the target dose. However, such practices pose the risk of inaccurate dosing, dose variation [31, 32] and dose-dumping for enteric coated tablets which could have severe therapeutic consequences [33, 34]. In addition, extemporaneous preparation can often be a particularly laborious and time-consuming task for clinical staff.

In these cases, 3DP could instead be used to create a printlet containing an exact dosage of drug. For example, 3DP has been used to print exact dosages of theophylline, a narrow therapeutic index (TI) drug used to treat asthma [35, 36]. In turn, this could simplify administration and reduce the risk of dose variation and medication errors. 3DP could also be used to simplify drug administration for those medicines that have rapid dose changes on initiation or reducing regimes. For example, corticosteroids can have complex reducing regimes and, as such, 3DP has been explored to dispense exact doses of prednisolone [37] and budesonide [35, 38].

With the advancement of genetic testing, future applications of 3DP could be used to print dosages based on a patient's genetic profile, metabolic pathways and biochemical markers. Their individual medicine and dosage could then be dispensed directly at the point of care, which would be a current challenge for centralised manufacturing processes.

2.2 *Multi-drug Combinations*

With an aging population, polypharmacy (the concurrent use of more than one medicine) is an increasing consideration that can cause non-adherence [39]. Patients who are hospitalised or have conditions such as tuberculosis or HIV can also be affected by a high tablet burden and complex dosage regimes [40]. In these instances,

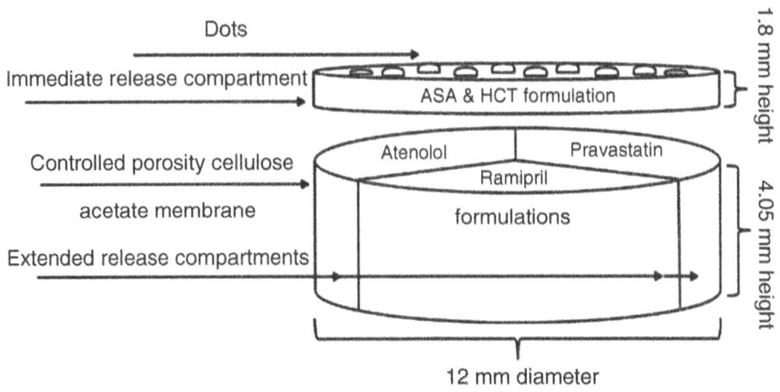

Fig. 1.2 3D printed polypill containing five different drugs with different release profiles. (Reprinted with permission from source: [44])

combining multiple actives and/or dosages into a single formulation (to create personalised 'polypills') could improve medication adherence and reduce administration errors. However, traditional manufacturing processes currently do not support true individualisation of polypills as they can only produce fixed-dose combinations that do not account for the changing needs of the patient.

Due to the excellent flexibility and capabilities for accurate spatial distribution of drug, 3DP could be used to produce polypills. Thus far, a number of 3DP papers have demonstrated the production of polypills using a range of printing technologies [41–43]. For example, Khaled et al. [44] successfully printed a five-drug cardiac polypill with different release rates, containing aspirin, hydrochlorothiazide, atenolol, ramipril and pravastatin (Fig. 1.2).

2.3 *Tailored Release Profiles*

Encouragingly, by selecting appropriate excipients and printing parameters, printlets could also be developed to have defined drug release profiles (e.g. by using different rate-controlling polymers). Several studies have evaluated this concept for single drugs [45–57] and polypills [41, 43, 44]. For example, one study described the production of enteric-coated caplets by printing a methacrylic polymer-based shell surrounding a drug core [35]. In 2017, Goyanes et al. [47] showed that delayed release tablets could instead be produced by incorporating the drug within an enteric polymer, negating the need for an outer enteric shell.

Release profiles could also be tailored by changing the geometry of the tablet [55, 58–60]. One study found that by changing the surface area to volume ratios of different shaped printlets (cube, pyramid, cylinder, sphere and torus; Fig. 1.3a) enabled the modification of drug release. When tablets were formulated with a constant surface area/volume ratio, the release rates (fastest first) were in the order of sphere and cube > torus > cylinder > pyramid (Fig. 1.3b).

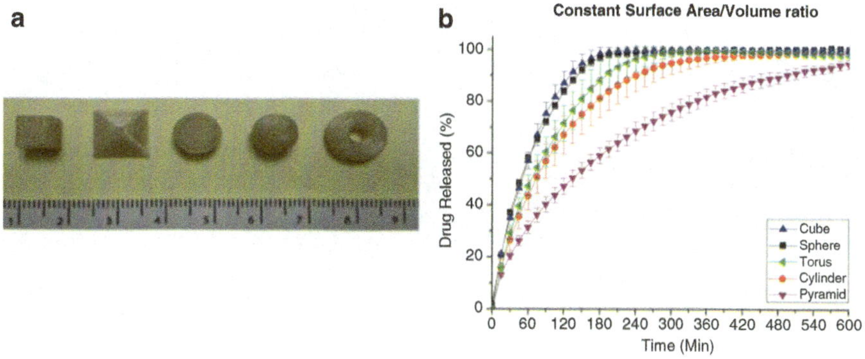

Fig. 1.3 (**a**) Printlets of different geometries (**b**) Drug release profiles of printlets with constant surface area/volume ratios. (Reprinted with permission from source: [46])

Moreover, novel strategies such as four-dimensional (4D) printing could provide excellent promise within the pharmaceutical sector, especially for the advancement of controlled drug delivery. 4D printing involves the fabrication of 3D objects that evolve to a pre-determined manner under the influence of external stimuli. As the goal of 4D printed objects are to alter their shape and size from external stimulation, the printed material (otherwise known as smart materials) used is required to be responsive to stimuli varying from heat, pH, magnetic field or light. The applications of 4D printing in healthcare and pharmaceuticals are discussed in further detail in Chap. 8.

2.4 Improved Patient Acceptability

Printing technologies can be used to produce 'patient-friendly' formulations. For example, the oral dosage form could be tailored to the patient's needs by enabling clinicians to select a specific required dosage and/or drug combinations. This benefit would improve treatment efficacy whilst reducing the risk of adverse effects due to inaccurate dosing [61]. Furthermore, patients or carers could be empowered to design or select their own formulation from a 'menu', choosing their desired formulation type (e.g. solid, chewable or oro-dispersible) and geometric factors such as shape, size, colour and texture, potentially improving medication adherence and understanding. Patient-friendly formulations, such as fast dissolving tablets [62] and oro-dispersible films [63–65] have been produced by 3DP. In both cases, these formulations could ease administration in patients with dysphagia, commonly demonstrated in paediatric and geriatric populations.

Using 3DP, formulations that would be otherwise impossible to produce with conventional mass manufacturing processes can be designed and printed. Goyanes et al. [66] evaluated patient acceptability (with regards to ease of swallowability and

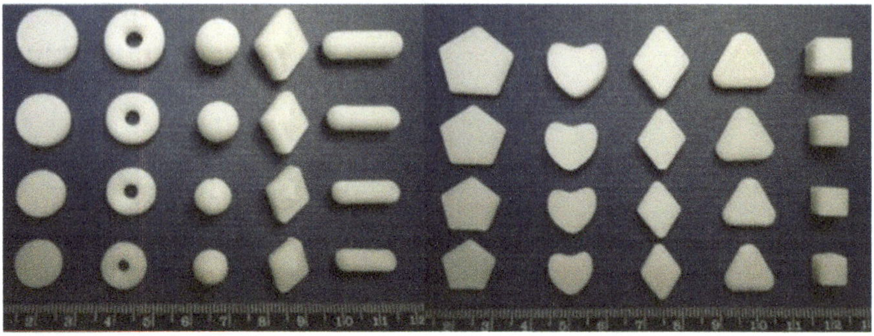

Fig. 1.4 Range of shapes and sizes of 3D printed tablets for evaluation of patient acceptability. (Reprinted with permission from source: [66])

ease of picking) of a variety of printlet shapes, sizes and colours (Fig. 1.4). It was found that the highest acceptability score relating to swallowing and picking was achieved with torus-shaped tablets. Printlets with a similar appearance to conventional formulations (e.g. caplets and discs) were also favoured, demonstrating that familiarity is key in medicines acceptance. This was the first study to demonstrate how 3D printed formulations could be tailored to the patient's preferences which could aid in increasing medication adherence.

2.5 *On-demand Printing*

As 3DP enables immediate manufacture, this technology is well suited to formulate those medicines with poor stability and hence short shelf lives on-demand. Examples include certain antibiotics post-reconstitution or those that require a desiccant (e.g. nitroglycerin for angina pectoris [67]). This was demonstrated in one study which printed exact dosages of dipyridamole [36] which can have an inherently poor shelf life (up to 6 weeks) once opened [68].

Due to on-demand capabilities, 3DP could be used across a number of healthcare services ranging from primary and secondary care (e.g. community pharmacies or specialist clinics) through to tertiary care (e.g. on hospital wards). On-demand dispensing in these settings could improve medicines access and accelerate discharge times, especially for medicines that require special's manufacture or extemporaneous preparation. Rapid production of one-off doses could even be achieved in time- or resource-constrained settings by integrating 3D printers into disaster areas, accident & emergency departments and first response units [69]. Areas with poor medicines access could also benefit from 3DP such as in military operations, third-world countries and even in space.

3DP could easily be integrated with other technological advancements of our generation such as smart health monitors, applications and cloud-based computing. In the twenty-first century, it is now feasible for patients to monitor their own vital signs (such as heart rate and blood pressure) using freely accessible applications on their smart phone devices. In the future, the clinician could access this real-time data within the existing internet infrastructure (termed Internet of Things; IoT), enabling easy review and modification of treatments or dosages. As 3D printers can be remotely controlled, the physician or pharmacist could the design a printlet and send a prescription to a location with a 3D printer to be dispensed [70]. It could also be possible for experienced patients to have their own 3D printers for medicine dispensing to ultimately facilitate autonomy in the treatment pathway.

3 Early Phase Drug Development

Early phase drug development covers pre-clinical studies through to Phase I clinical trials. During these studies, a wide dose range is often administered to gain initial information around safety, tolerability, toxicity and safety (Fig. 1.5). The data collected often influences critical decision-making around accepting or rejecting new drugs to take forward into later phase trials. Nowadays, the rate of drug failure and cost of taking a new chemical entity to market is high during early phase development; in 2014 only 11% of drugs in Phase I trials were approved by the Food and Drug Administration (FDA) [71].

Appropriate formulation selection and optimisation plays a major part in the drug development timeline. The ideal formulation would be inexpensive, display adequate bioavailability, have high dose flexibility and facilitate rapid progression. However, traditional manufacturing processes often do not support such requirements. For example, whilst being well suited and cost effective for mass manufacture, procedures such as powder compaction are unsuitable for the smaller batch production required in early phase development. This is due to the processes being inherently dose inflexible, lengthy, labour intensive and expensive.

As such, it is evident that new approaches are required to produce customised dosage forms to increase efficiency and reduce cost. In turn, this could enable a rapid identification of suitable drugs at minimal cost early on in drug development, ideally within pre-clinical studies or FIH clinical trials. 3DP could prove advantageous for this application by enabling the production of small or 'one-off' batches of formulations in a cost-effective, efficient and flexible manner. Using such technology could expedite entry into clinical trials, and hence facilitate the rapid

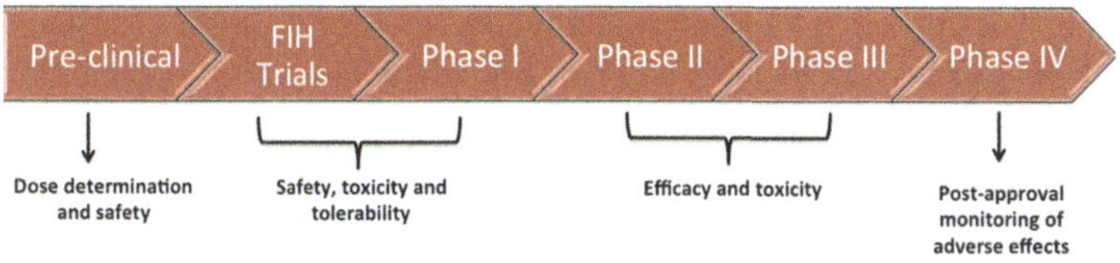

Fig. 1.5 Schematic of the drug development process

clinical go/no-go decisions required early on in drug development. The applications of 3DP for formulation production within early drug development will be discussed, covering both pre-clinical studies and FIH clinical trials.

3.1 Pre-clinical Studies and FIH Trials

Pre-clinical studies involve a series of *in vitro* and *in vivo* tests to gain an initial understanding of drug efficacy, toxicology, safety and pharmacokinetic (PK) behaviour [72]. The data obtained often paves the way in supporting the selection (or rejection) of new drug candidates, and associated dosing schedules, to take forward into FIH clinical trials [73, 74]. It is desirable to rapidly progress through pre-clinical work to enable quick FIH trial entry, as well as to limit time and resource investment.

In vivo studies typically involve administration of formulations to rodents such as rats as they are inexpensive, easy to reproduce and display physiological similarities with humans (e.g. intestinal absorption and gastric emptying) [75]. Typical formulations used in pre-clinical work include oral solutions and suspensions and small capsules/tablets, which are easy to administer.

Following pre-clinical studies, the drug can progress into FIH clinical trials, which represent the first administration of the drug into humans. These studies have the primary aim of evaluating the drug's safety, tolerability and pharmacokinetic behaviour in a small number of volunteers (typically 20–30) [76]. Administration in humans begins at low (and often heavily sub-therapeutic) doses based on pre-clinical toxicology data [74]. Doses are then escalated incrementally and often span a wide dose range; it has been estimated that dosing over an 800-fold range may be required [77, 78]. The most common FIH trial designs are single or multiple ascending dose designs (Fig. 1.6a, b) [79, 80]. In recent years, adoption of multi-part adaptive trial designs has enabled a more flexible approach to trial design [81, 82]. Compared to conventional strategies, adaptive approaches enable the pre-defined modification of certain aspects of the trial [83]. The flexible design uses real-time clinical data to guide future decisions on dosing, subject population, formulation selection and sample size [84].

It is evident that due to the frequent dose changes and variable nature of FIH trials, the selection of an appropriate formulation (e.g. with high dose flexibility, rapid manufacture and low cost) is of paramount importance. Conventionally, two main dosage forms are used: liquid formulations (ready-to-use solutions and suspensions or drug-in-bottle) and solid oral dosage forms (drug-in-capsule or formulated tablets and capsules) [85, 86]. However, these formulations come alongside a number of issues, ranging from drug solubility and stability (e.g. for liquid formulations) through to high resource investment and inadequate dose flexibility (e.g. for formulated tablets and capsules) [85, 86].

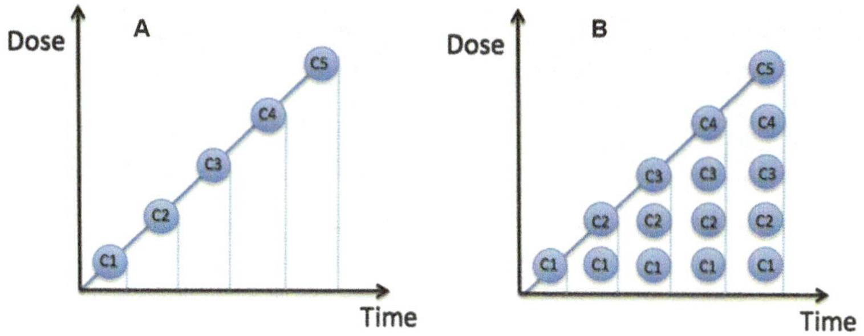

Fig. 1.6 Commonly used FIH trial designs including (**a**) single ascending dose designs and (**b**) multiple ascending dose designs. (**c**) represents the cohort number. The vertical dashed lines represent safety evaluation prior to subsequent dosing

3.2 *Motivations for Using 3DP in Early Phase Drug Development*

Adoption of 3DP as an alternative tablet production tool may provide many benefits to early phase drug development by enabling an inexpensive and rapid means of small batch production of highly dose flexible formulations. The main motivations of using 3DP in pre-clinical studies and FIH trials can be broken down into four main categories, which will be discussed in turn;

3.2.1 Dose Flexibility

3DP is a highly flexible process which permits easy modification of dosages suited to the needs of the pre-clinical study or FIH trial. This is favourable as frequent dose escalations are common during early drug development. During pre-clinical studies, doses are commonly altered ‘real-time’ based on toxicological and pharmacological results. Dose flexibility is typically achieved using liquid formulations, although concerns around drug solubility, stability and taste may limit its use. Formulated tablets and capsules may also be used, however require lengthy manufacturing processes and are produced in limited, discrete strengths. Other formulations, such as drug-in-capsule and drug-in-bottle, require staff to manually dispense the exact dose, which can be laborious and time-consuming [85, 86].

3DP could overcome such challenges by dispensing small batches of printlets containing a multitude of dosages. This can be achieved in a number of ways. First, by physically modifying the tablet dimensions or infill percentage (the degree to which the internal space will be filled from 0%, hollow, to 100%, solid) a greater or smaller amount of material, and hence dose, can be deposited [87]. Changing the physical dimensions enabled one study to create printlets covering a wide dose range (60 – 300 mg) (Fig. 1.7) [88]. Second, by producing ‘feedstock’ with different initial drug loadings could change the drug concentration in the final dosage form.

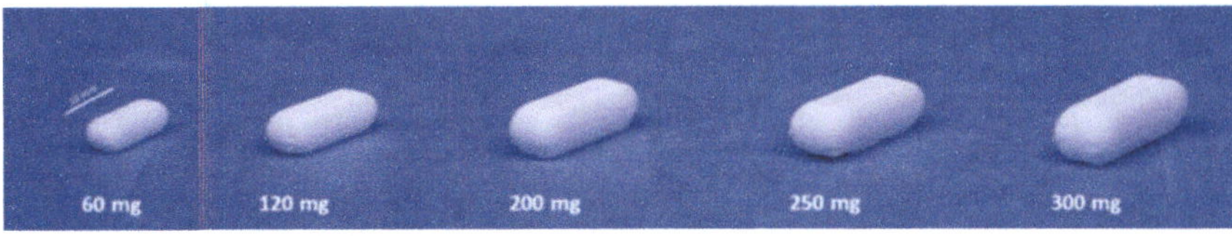

Fig. 1.7 Dose variation by modification of printlet dimensions. (Reprinted with permission from source: [88])

Within pre-clinical work, the research scientist could use 3DP to produce small printlets containing exact dosages designed to suit the study and animal requirements. This could provide an easier, more efficient and more precise means of dose evaluation and data collection, potentially expediting FIH trial entry. Within FIH trials, an improvement in participant convenience, reduction in resource investment and increase in efficiency may be achieved by limiting the number of tablet strengths, or the volume of liquid required to attain the target dose. This may be particularly advantageous if testing a wide dose range and especially for the newer multi-part adaptive trial designs, which use real-time PK and safety data to guide future FIH trial dosing decisions [84].

3.3 *Immediate Manufacture*

3D printers are small, compact systems that can be easily integrated into a laboratory or clinical trial setting. This technology can produce printlets rapidly (in seconds) which is an attractive concept for resource- or time-constrained settings, such as within pre-clinical formulation development. On demand printing enables the rapid production of several product iterations for testing which would enable formulators to assess factors such as suitable excipient inclusion and compatibility. Compared to traditional manufacturing processes, development of an optimal product could occur swiftly without increasing lead-times or development costs [89].

Within FIH trials, 3DP could also be highly beneficial for medicines with inherently poor stability such as liquid formulations. Producing printlets immediately pre-dosing could reduce the length of storage as well as prevent the need for stability-improving measures (e.g. refrigeration or addition of preservatives). This could negate the need for long-term accelerated storage testing which typically takes 4 weeks for probe batches; a condition that often delays trial entry [80]. Short-term stability testing may still be required to ensure drug stability post-dispensing, however this would only require short evaluation (e.g. 7 days).

3.4 Reduced Resource Investment

Conventional formulations used in drug development require time- and resource-intensive manufacturing processes. For example, within FIH trials, a common strategy is to use a drug-in-bottle or drug-in-capsule formulation for reconstitution on site [79]. In many cases the containers or capsules require manual filling which can be particularly laborious for trial staff. Other traditional manufacturing processes, such as powder compaction, require large amounts of starting materials, lengthy multi-step processes (e.g. milling, mixing, granulation etc.) and require major capital input. These large batch processes also require investment in several pieces of expensive equipment, large working spaces and highly skilled staff for efficient operation.

Using 3DP as an alternative dispensing tool could alleviate these manufacturing concerns. 3D printers have a low cost and are small, compact systems that require little operation space. Favourably, 3DP can produce entire dosage forms in a simple, repeatable and reproducible unit operation. By permitting exact material deposition, on-demand manufacture using 3DP is a much more drug-sparing process, requiring lower amounts of starting ingredients whilst reducing wastage. 3D printers are also very simple to use and do not have a high training burden.

3.5 Unique Characteristics

3DP can be used to create simple and complex printlets with unique characteristics. Unlike traditional manufacturing processes, an accurate spatial distribution of drug and excipients can be attained, opening up the potential for a vast array of formulation designs and geometries to be produced [90, 91]. This could be exploited in a number of cases within pre-clinical development and FIH trials.

By integrating a 3D printer into a laboratory setting, the scientist undertaking pre-clinical work could customise printlets based on the pharmacokinetic and pharmacodynamic characteristics of the animal model. For example, small pellets, mini-tablets, small capsules (size 9) and smaller devices are often used as a convenient means of dosing rodents in pre-clinical studies [92]. A study by Goyanes et al. demonstrated that 3D printed device design could be used to fabricate dosage forms of varying sizes and geometries with better gastric emptying characteristics suitable for rodent administration [93].

For poorly soluble compounds (e.g. Biopharmaceutical Classification System; BCS Class II and IV drugs), the use of enabling strategies is sometimes warranted to improve bioavailability and hence reduce the amount of drug required for therapeutic effects [94]. A commonly used strategy is to create a solid dispersion, whereby the amorphous form of the drug is stabilised within a polymer matrix, for example by spray drying [95] or using hot melt extrusion (HME) [96, 97]. In this respect, thermal 3DP methods coupled with HME could be exploited as a potential

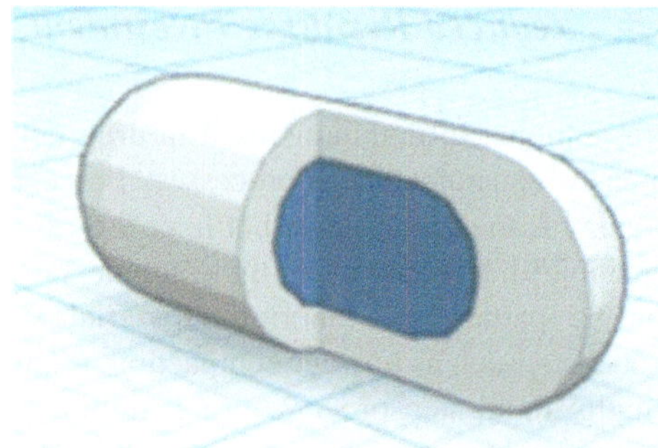

Fig. 1.8 DuoCaplet design. (Reprinted with permission from [99])

enabling strategy [98]. As such, with a greater number of low solubility compounds entering drug development, 3DP could provide a useful enabling strategy to aid formulation development.

Within FIH trials, printlets could also be tailored to ease administration, in turn increasing patient recruitment and reducing dropout rates. This could be particularly useful for unblinded trials (such as within oncology trials [99]) for those patients with swallowing difficulties that might otherwise be excluded (e.g. paediatrics and geriatric populations). If blinding is required (a process that prevents the participant from knowing which medicine they are receiving [100]), printlets could also be designed to mask the presence of the drug. For example, a two-compartment device could be printed, comprised of a drug 'core' embedded within a larger inert polymer tablet. This theory was demonstrated within a study which printed DuoCaplets whereby a paracetamol core was imbedded within a larger caffeine caplet (Fig. 1.8) [41]. Using a DuoCaplet design could be useful if the drug is coloured (and hence is noticeably different to the placebo), for taste-masking or to avoid the knowledge of dose administration (e.g. by changing the size of the core whilst maintaining the overall tablet size).

4 Conclusion

The numerous applications of 3D printing to pharmaceuticals are evident. This technology could revolutionise formulation production, to move away from mass manufacture to producing highly flexible and customised dosage forms on-demand. Tailoring formulations in this way could largely drive the field of personalised medicine, as well as benefit early phase drug development, including pre-clinical studies and FIH trials. The recent FDA approval of the first 3D printed tablet (Spritam®) was a significant milestone in this technology's history. 3DP is expected to continue rapidly evolving over the next decade. Progress is already being made, with cutting-edge pharmaceutical research being published every day, detailing the new possibilities that 3DP can bring.

References

1. Morgan J. What is the fourth industrial revolution? Forbes. 2016.; Available from: https://www.forbes.com/sites/jacobmorgan/2016/02/19/what-is-the-4th-industrialrevolution/#65e28c00f392
2. Chowdhry A. What can 3D printing do? Here are 6 creative examples. Forbes. 2013.; Available from: https://www.forbes.com/sites/amitchowdhry/2013/10/08/what-can-3dprinting- do-here-are-6-creative-examples/
3. Sahlgren C, Meinander A, Zhang H, Cheng F, Preis M, Xu C, Salminen TA, Toivola D, Abankwa D, Rosling A, Şen Karaman D, Salo-Ahen OMH, Österbacka R, Eriksson JE, Willför S, Petre I, Peltonen J, Leino R, Johnson M, Rosenholm J, Sandler N. Tailored approaches in drug development and diagnostics: from molecular design to biological model systems. Adv Healthc Mater. 2017;6(21):1700258.
4. Miller JS, Stevens KR, Yang MT, Baker BM, Nguyen DH, Cohen DM, et al. Rapid casting of patterned vascular networks for perfusable engineered three-dimensional tissues. Nat Mater. 2012;11(9):768–74.
5. Leukers B, Gulkan H, Irsen SH, Milz S, Tille C, Schieker M, et al. Hydroxyapatite scaffolds for bone tissue engineering made by 3D printing. J Mater Sci Mater Med. 2005;16(12):1121–4.
6. Ventola CL. Medical applications for 3D printing: current and projected uses. Pharmacy and Therapeutics. 2014;39(10):704–11.
7. Aprecia Pharmaceutials. 3D printing - ZipDose technology. 2015. Available from: https://aprecia.com/zipdose-platform/3d-printing.php.
8. Muwaffak Z, Goyanes A, Clark V, Basit AW, Hilton ST, Gaisford S. Patient-specific 3D scanned and 3D printed antimicrobial polycaprolactone wound dressings. Int J Pharm. 2017;527(1–2):161–70.
9. Goyanes A, Det-Amornrat U, Wang J, Basit AW, Gaisford S. 3D scanning and 3D printing as innovative technologies for fabricating personalized topical drug delivery systems. J Control Release. 2016;234:41–8.
10. Waran V, Pancharatnam D, Thambinayagam HC, Raman R, Rathinam AK, Balakrishnan YK, et al. The utilization of cranial models created using rapid prototyping techniques in the development of models for navigation training. J Neurol Surg A Cent Eur Neurosurg. 2014;75(1):12–5.
11. Li J, Nie L, Li Z, Lin L, Tang L, Ouyang J. Maximizing modern distribution of complex anatomical spatial information: 3D reconstruction and rapid prototype production of anatomical corrosion casts of human specimens. Anat Sci Educ. 2012;5(6):330–9.
12. Nakada T, Akiba T, Inagaki T, Morikawa T. Thoracoscopic anatomical subsegmentectomy of the right S2b + S3 using a 3D printing model with rapid prototyping. Interact Cardiovasc Thorac Surg. 2014;19(4):696–8.
13. Rengier F, Mehndiratta A, von Tengg-Kobligk H, Zechmann CM, Unterhinninghofen R, Kauczor HU, et al. 3D printing based on imaging data: review of medical applications. Int J Comput Assist Radiol Surg. 2010;5(4):335–41.
14. Mendoza HR. Researches study the 3D printing of surgical instruments for use in long space missions. 3DPrint.com. 2014. Available from: https://3dprint.com/8937/3dprinting-space-missions/.
15. Saunders S. Wrist osteotomy guide. 3DPrint.Com. 2017. Available from: https://www.3dprint.com/165853/materialise-mimics-wrist-osteotomy/wrist-osteotomy-guide/.
16. Wyman C. Some straight talk (and a Round Award) for 3D printing in orthodontics. Stratasys. 2014. Available from: http://blog.stratasys.com/2014/02/28/dental-orthodontics- 3d-printer/.
17. Trenfield SJ, Awad A, Goyanes A, Gaisford S, Basit AW. 3D printing pharmaceuticals: drug development to front-line care. Trends Pharmacol Sci. 2018; 39(5):440–451.
18. Peng W, Datta P, Ayan B, Ozbolat V, Sosnoski D, Ozbolat IT. 3D bioprinting for drug discovery and development in pharmaceutics. Acta Biomater. 2017;57:26–46.

19. Collins FS, Varmus H. A new initiative on precision medicine. N Engl J Med. 2015;372(9):793–5.
20. Alomari M, Mohamed FH, Basit AW, Gaisford S. Personalised dosing: printing a dose of one's own medicine. Int J Pharm. 2015;494(2):568–77.
21. Freire AC, Basit AW, Choudhary R, Piong CW, Merchant HA. Does sex matter? The influence of gender on gastrointestinal physiology and drug delivery. Int J Pharm. 2011;415(1–2):15–28.
22. Merchant HA, Liu F, Orlu Gul M, Basit AW. Age-mediated changes in the gastrointestinal tract. Int J Pharm 2016;512(2):382–95.
23. Hamburg MA, Collins FS. The path to personalized medicine. N Engl J Med. 2010;363(4):301–4.
24. Hatton GB, Madla CM, Rabbie SC, Basit AW. All disease begins in the gut: Influence of gastrointestinal disorders and surgery on oral drug performance. Int J Pharm. 2018;548(1):408–22.
25. Yadav V, Varum F, Bravo R, Furrer E, Bojic D, Basit AW. Inflammatory bowel disease: exploring gut physiology for novel therapeutic targets. Transl Res. 2016;176:38–68.
26. Taherali F, Varum F, Basit AW. A slippery slope: on the origin, role and physiology of mucus. Adv Drug Deliv Rev. 2018;124:16–33.
27. FDA. Paving the way for personalized medicine. 2013. Available from: https://www.fda.gov/downloads/ScienceResearch/SpecialTopics/PersonalizedMedicine/UCM372421.pdf.
28. Awad A, Trenfield SJ, Goyanes A, Gaisford S, Basit AW. Reshaping drug development using 3D printing. Drug Discov Today. 2018; https://doi.org/10.1016/j.drudis.2018.05.025.
29. Kearns GL, Abdel-Rahman SM, Alander SW, Blowey DL, Leeder JS, Kauffman RE. Developmental pharmacology--drug disposition, action, and therapy in infants and children. N Engl J Med. 2003;349(12):1157–67.
30. Breitkreutz J, Boos J. Paediatric and geriatric drug delivery. Expert Opin Drug Deliv. 2007;4(1):37–45.
31. Peek BT, Al-Achi A, Coombs SJ. Accuracy of tablet splitting by elderly patients. JAMA. 2002;288(4):451–2.
32. Habib WA, Alanizi AS, Abdelhamid MM, Alanizi FK. Accuracy of tablet splitting: comparison study between hand splitting and tablet cutter. Saudi Pharm J. 2014;22(5):454–9.
33. McDevitt JT, Gurst AH, Chen Y. Accuracy of tablet splitting. Pharmacother: J Hum Pharmacol Drug Ther. 1998;18(1):193–7.
34. Hill S, Varker AS, Karlage K, Myrdal PB. Analysis of drug content and weight uniformity for half-tablets of 6 commonly split medications. J Manag Care Pharm. 2009;15(3):253–61.
35. Okwuosa TC, Pereira BC, Arafat B, Cieszynska M, Isreb A, Alhnan MA. Fabricating a shell-core delayed release tablet using dual FDM 3D printing for patient-centred therapy. Pharm Res. 2017;34(2):427–37.
36. Okwuosa TC, Stefaniak D, Arafat B, Isreb A, Wan K-W, Alhnan MA. A lower temperature FDM 3D printing for the manufacture of patient-specific immediate release tablets. Pharm Res. 2016;33(11):2704–12.
37. Skowyra J, Pietrzak K, Alhnan MA. Fabrication of extended-release patient-tailored prednisolone tablets via fused deposition modelling (FDM) 3D printing. Eur J Pharm Sci. 2015;68:11–7.
38. Goyanes A, Chang H, Sedough D, Hatton GB, Wang J, Buanz A, et al. Fabrication of controlled-release budesonide tablets via desktop (FDM) 3D printing. Int J Pharm. 2015;496(2):414–20.
39. Maher RL, Hanlon J, Hajjar ER. Clinical consequences of polypharmacy in elderly. Expert Opin Drug Saf. 2014;13(1):57–65.
40. Murray MD, Kroenke K. Polypharmacy and medication adherence: small steps on a long road. J Gen Intern Med. 2001;16(2):137–9.
41. Goyanes A, Wang J, Buanz A, Martinez-Pacheco R, Telford R, Gaisford S, et al. 3D printing of medicines: engineering novel oral devices with unique design and drug release characteristics. Mol Pharm. 2015;12(11):4077–84.
42. Rowe CW, Katstra WE, Palazzolo RD, Giritlioglu B, Teung P, Cima MJ. Multimechanism oral dosage forms fabricated by three dimensional printing™. J Control Release. 2000;66(1):11–7.

43. Khaled SA, Burley JC, Alexander MR, Yang J, Roberts CJ. 3D printing of tablets containing multiple drugs with defined release profiles. Int J Pharm. 2015;494(2):643–50.
44. Khaled SA, Burley JC, Alexander MR, Yang J, Roberts CJ. 3D printing of five-in-one dose combination polypill with defined immediate and sustained release profiles. J Control Release. 2015;217:308–14.
45. Gioumouxouzis CI, Katsamenis OL, Bouropoulos N, Fatouros DG. 3D printed oral solid dosage forms containing hydrochlorothiazide for controlled drug delivery. J Drug Deliv Sci Technol. 2017;40:164–71.
46. Goyanes A, Buanz ABM, Hatton GB, Gaisford S, Basit AW. 3D printing of modified-release aminosalicylate (4-ASA and 5-ASA) tablets. Eur J Pharm Biopharm. 2015;89:157–62.
47. Goyanes A, Fina F, Martorana A, Sedough D, Gaisford S, Basit AW. Development of modified release 3D printed tablets (printlets) with pharmaceutical excipients using additive manufacturing. Int J Pharm. 2017;527(1–2):21–30.
48. Fina F, Madla CM, Goyanes A, Zhang J, Gaisford S, Basit AW. Fabricating 3D printed orally disintegrating printlets using selective laser sintering. Int J Pharm. 2018;541(1–2):101–7.
49. Martinez PR, Goyanes A, Basit AW, Gaisford S. Fabrication of drug-loaded hydrogels with stereolithographic 3D printing. Int J Pharm. 2017;532(1):313–7.
50. Beck RCR, Chaves PS, Goyanes A, Vukosavljevic B, Buanz A, Windbergs M, Basit AW, Gaisford S. 3D printed tablets loaded with polymeric nanocapsules: an innovative approach to produce customized drug delivery systems. Int J Pharm. 2017;528(1–2):268–79.
51. Goyanes A, Kobayashi M, Martínez-Pacheco R, Gaisford S, Basit AW. Fused-filament 3D printing of drug products: microstructure analysis and drug release characteristics of PVA-based caplets. Int J Pharm. 2016;514(1):290–5.
52. Wang J, Goyanes A, Gaisford S, Basit AW. Stereolithographic (SLA) 3D printing of oral modified-release dosage forms. Int J Pharm. 2016;503(1–2):207–12.
53. Kollamaram G, Croker DM, Walker GM, Goyanes A, Basit AW, Gaisford S. Low temperature fused deposition modeling (FDM) 3D printing of thermolabile drugs. Int J Pharm. 2018;545(1–2):144–52.
54. Sadia M, Arafat B, Ahmed W, Forbes RT, Alhnan MA. Channelled tablets: an innovative approach to accelerating drug release from 3D printed tablets. J Control Release. 2018;269:355–63.
55. Martinez PR, Goyanes A, Basit AW, Gaisford S. Influence of geometry on the drug release profiles of stereolithographic (SLA) 3D printed tablets. AAPS PharmSciTech. 2018; https://doi.org/10.1208/s12249-018-1075-3.
56. Maroni A, Melocchi A, Parietti F, Foppoli A, Zema L, Gazzaniga A. 3D printed multi-compartment capsular devices for two-pulse oral drug delivery. J Control Release. 2017;268:10–8.
57. Solanki NG, Tahsin M, Shah AW, Serajuddin ATM. Formulation of 3D printed tablet for rapid drug release by fused deposition modeling: screening polymers for drug release, drug-polymer miscibility and printability. J Pharm Sci. 2018;107(1):390–401.
58. Fina F, Goyanes A, Madla CM, Awad A, Trenfield SJ, Kuek JM, Patel P, Gaisford S, Basit AW. 3D printing of drug-loaded gyroid lattices using selective laser sintering. Int J Pharm. 2018; https://doi.org/10.1016/j.ijpharm.2018.05.044.
59. Goyanes A, Robles Martinez P, Buanz A, Basit AW, Gaisford S. Effect of geometry on drug release from 3D printed tablets. Int J Pharm. 2015;494(2):657–63.
60. Fina F, Goyanes A, Gaisford S, Basit AW. Selective laser sintering (SLS) 3D printing of medicines. Int J Pharm. 2017;529(1-2):285–93.
61. Diment LE, Thompson MS, Bergmann JHM. Clinical efficacy and effectiveness of 3D printing: a systematic review. BMJ Open. 2017;7(12):e016891.
62. Yu D-G, Branford-White C, Yang Y-C, Zhu L-M, Welbeck EW, Yang X-L. A novel fast disintegrating tablet fabricated by three-dimensional printing. Drug Dev Ind Pharm. 2009;35(12):1530–6.

63. Buanz AB, Saunders MH, Basit AW, Gaisford S. Preparation of personalized-dose salbutamol sulphate oral films with thermal ink-jet printing. Pharm Res. 2011;28(10):2386–92.
64. Jamróz W, Kurek M, Łyszczarz E, Szafraniec J, Knapik-Kowalczuk J, Syrek K, et al. 3D printed orodispersible films with Aripiprazole. Int J Pharm. 2017;533:413–20.
65. Vuddanda PR, Alomari M, Dodoo CC, Trenfield SJ, Velga S, Basit AW, Gaisford S. Personalisation of warfarin therapy using thermal ink-jet printing. Eur J Pharm Sci. 2018;117:80–7.
66. Goyanes A, Scarpa M, Kamlow M, Gaisford S, Basit AW, Orlu M. Patient acceptability of 3D printed medicines. Int J Pharm. 2017;530(1):71–8.
67. Kommanaboyina B, Rhodes CT. Trends in stability testing, with emphasis on stability during distribution and storage. Drug Dev Ind Pharm. 1999;25(7):857–68.
68. eMC. Persantin retard 200 mg (Dipyridamole). 2015. Available from: https://www.medicines.org.uk/emc/medicine/304-SHELF_LIFE.
69. Norman J, Madurawe RD, Moore CM, Khan MA, Khairuzzaman A. A new chapter in pharmaceutical manufacturing: 3D-printed drug products. Adv Drug Deliv Rev. 2017;108:39–50.
70. Alhnan MA, Okwuosa TC, Sadia M, Wan KW, Ahmed W, Arafat B. Emergence of 3D printed dosage forms: opportunities and challenges. Pharm Res. 2016;33(8):1817–32.
71. Hay M, Thomas DW, Craighead JL, Economides C, Rosenthal J. Clinical development success rates for investigational drugs. Nat Biotechnol. 2014;32(1):40–51.
72. Gao Y, Gesenberg C, Zheng W. Chapter 17 – Oral formulations for preclinical studies: principle, design, and development considerations A2 – Qiu, Yihong. In: Chen Y, Zhang GGZ, Yu L, Mantri RV, editors. Developing solid oral dosage forms. 2nd ed. Boston: Academic Press; 2017. p. 455–95.
73. Dorato MA, Engelhardt JA. The no-observed-adverse-effect-level in drug safety evaluations: use, issues, and definition(s). Regul Toxicol Pharmacol. 2005;42(3):265–74.
74. FDA. Estimating the maximum safe starting dose in initial clinical trials for therapeutics in adult healthy volunteers. 2005. Available from: https://www.fda.gov/downloads/drugs/guidances/ucm078932.pdf.
75. Hatton GB, Yadav V, Basit AW, Merchant HA. Animal farm: considerations in animal gastrointestinal physiology and relevance to drug delivery in humans. J Pharm Sci. 2015;104(9):2747–76.
76. Buoen C, Bjerrum OJ, Thomsen MS. How first-time-in-human studies are being performed: a survey of phase I dose-escalation trials in healthy volunteers published between 1995 and 2004. J Clin Pharmacol. 2005;45(10):1123–36.
77. Ku MS, Dulin W. A biopharmaceutical classification-based right-first-time formulation approach to reduce human pharmacokinetic variability and project cycle time from first-in-human to clinical proof-of-concept. Pharm Dev Technol. 2012;17(3):285–302.
78. Tong G, Wang J-S, Sverdlov O, Huang S-P, Slemmon R, Croop R, et al. Multicenter, randomized, double-blind, placebo-controlled, single-ascending dose study of the oral γ-secretase inhibitor BMS-708163 (Avagacestat): tolerability profile, pharmacokinetic parameters, and pharmacodynamic markers. Clin Ther. 2012;34(3):654–67.
79. ABPI. First in human studies: points to consider in study placement, design and conduct. 2011. Available from: http://www.abpi.org.uk/ourwork/ library/guidelines/Documents/First in Human Studies.pdf.
80. Capretto L, Byrne G, Trenfield SJ, Dowden L, Booth S. Formulation, analytical and regulatory strategies for first-in-human clinical trials. Oral formulation roadmap from early drug discovery to development. Hoboken: Wiley; 2017. p. 165–241.
81. Mahajan R, Gupta K. Adaptive design clinical trials: methodology, challenges and prospect. Indian J Pharm. 2010;42(4):201–7.
82. Perlstein I, Bolognese JA, Krishna R, Wagner JA. Evaluation of agile designs in first-in-human (FIH) trials—a simulation study. AAPS J. 2009;11(4):653.
83. EUPATI. New approaches to clinical trials: adaptive designs. 2015. Available from: https://www.eupati.eu/clinical-development-and-trials/new-approaches-to-clinical-trialsadaptive- designs/-Adaptive_designs.

84. Brennan Z. CROs slowly shifting to adaptive clinical trial designs outsourcing pharma. 2013. Available from: http://www.outsourcing-pharma.com/Clinical-Development/CROs- Slowly-Shifting-to-Adaptive-Clinical-Trial-Designs.
85. Hariharan M, Ganorkar L, Amidon G, Cavallo A, Gatti P, Hageman MJ, et al. Reducing the time to develop and manufacture formulations for first oral dose in humans. Pharm Technol. 2003.; Available from: http://www.capsugel.com/media/library/reducing_the_time_to_develop_and_manufacture_formulations_for_first_oral_dose_in_humans.pdf
86. Kuentz M, Holm R, Elder DP. Methodology of oral formulation selection in the pharmaceutical industry. Eur J Pharm Sci. 2016;87:136–63.
87. Goyanes A, Buanz ABM, Basit AW, Gaisford S. Fused-filament 3D printing (3DP) for fabrication of tablets. Int J Pharm. 2014;476(1):88–92.
88. Pietrzak K, Isreb A, Alhnan MA. A flexible-dose dispenser for immediate and extended release 3D printed tablets. Eur J Pharm Biopharm. 2015;96:380–7.
89. Stratasys. How 3D printing will continue to transform manufacturing. Stratasys Direct. 2015. Available from: https://www.stratasysdirect.com/content/white_papers/str_7463_15_sdm_wp_transform_mfg.pdf.
90. Awad A, Trenfield SJ, Gaisford S, Basit AW. 3D printed medicines: A new branch of digital healthcare. Int J Pharm. 2018;548(1):586–96.
91. Zema L, Melocchi A, Maroni A, Gazzaniga A. Three-dimensional printing of medicinal products and the challenge of personalized therapy. J Pharm Sci. 2017;106(7):1697–705.
92. Shah AK, Agnihotri SA. Recent advances and novel strategies in pre-clinical formulation development: an overview. J Control Release. 2011;156(3):281–96.
93. Goyanes A, Fernández-Ferreiro A, Majeed A, Gomez-Lado N, Awad A, Luaces-Rodríguez A, Gaisford S, Aguiar P, Basit AW. PET/CT imaging of 3D printed devices in the gastrointestinal tract of rodents. Int J Pharm. 2018;536(1):158–64.
94. Ayad MH. Rational formulation strategy from drug discovery profiling to human proof of concept. Drug Deliv. 2015;22(6):877–84.
95. Dobry DE, Settell DM, Baumann JM, Ray RJ, Graham LJ, Beyerinck RA. A model-based methodology for spray-drying process development. J Pharm Innov. 2009;4(3):133–42.
96. Breitenbach J. Melt extrusion: from process to drug delivery technology. Eur J Pharm Biopharm. 2002;54(2):107–17.
97. Tiwari RV, Patil H, Repka MA. Contribution of hot-melt extrusion technology to advance drug delivery in the 21st century. Expert Opin Drug Deliv. 2016;13(3):451–64.
98. Zhang J, Feng X, Patil H, Tiwari RV, Repka MA. Coupling 3D printing with hot-melt extrusion to produce controlled-release tablets. Int J Pharm. 2017;519(1–2):186–97.
99. Le Tourneau C, Lee JJ, Siu LL. Dose escalation methods in phase I cancer clinical trials. JNCI J Natl Canc Inst. 2009;101(10):708–20.
100. Page SJ, Persch AC. Recruitment, retention, and blinding in clinical trials. Am J Occup Ther. 2013;67(2):154–61.

Chapter 2
3D Printing Technologies, Implementation and Regulation: An Overview

Christine M. Madla, Sarah J. Trenfield, Alvaro Goyanes, Simon Gaisford, and Abdul W. Basit

Abstract The rise in three-dimensional (3D) printing in design and manufacturing, like any other, is the product of vision and implementation, pioneered by those who were brave enough to make it happen. In this chapter, the advancements of exponential developments driven by 3D printers themselves and its application in almost all areas of manufacturing and personalisation, namely; aeronautics, engineering, architecture and pharmaceutics are discussed. This chapter further serves to provide an introduction to the different 3D printing technologies, their respective histories, potential benefits, limitations and regulatory requirements, and a thorough description of the new and exciting possibilities that can arise by simply acknowledging the capabilities of 3D printing in healthcare.

Keywords Additive manufacturing · Medical devices · Drug delivery systems · Personalized pharmaceuticals · Digital pharmacy · Bioprinting

1 Introduction

We are living in a technological era; the present day has been described as the "next industrial revolution" [1, 2] where living in a world of cloud-based computing, contactless payments, artificial intelligence and three-dimensional (3D) printing,

C. M. Madla (✉) · S. J. Trenfield
Department of Pharmaceutics, UCL School of Pharmacy,
University College London, London, UK
e-mail: christine.madla.16@ucl.ac.uk

A. Goyanes
FabRx Ltd., Ashford, Kent, UK

S. Gaisford · A. W. Basit
Department of Pharmaceutics, UCL School of Pharmacy,
University College London, London, UK

FabRx Ltd., Ashford, Kent, UK

A. W. Basit, S. Gaisford (eds.), *3D Printing of Pharmaceuticals*,
AAPS Advances in the Pharmaceutical Sciences Series 31,
https://doi.org/10.1007/978-3-319-90755-0_2

which were once a fantasy, have now become a reality. It is important, now more than ever, for the pharmaceutical industry to not resist change but to instead, understand and embrace the power of innovation to support manufacturing processes such as analytical technologies and 3D printing. This chapter provides a brief history of 3D printing and holistically describes the functions of a portfolio of 3D printing techniques from; binder jet printing, stereolithography, selective laser sintering, fused deposition modeling and semi-solid gel extrusion, the fast-emerging four-dimensional (4D) printing and the applications of 3D printing in the scope of pharmaceutical science. A more thorough description of the independent histories, developments, applications, benefits and limitations of a multitude of 3D printing techniques are provided in the subsequent chapters to come.

2 Classification of 3D Printing Technologies

3D printing is generally used as synonym for additive manufacturing, solid-freeform technology (SFF) or rapid prototyping (RP). These terms are defined as the production of a prototype or a scale model of a physical object using 3D computer-aided design (CAD) software [3].

The first standards around terminology were developed by the American Society for Testing and Materials (ASTM) International, an international standards organisation that develops and publishes voluntary consensus technical standards for a wide range of materials, products and systems [3]. Originally, the term additive manufacturing was considered to be more inclusive for the industry. 3D printing in particular was associated with machines that were low in price and/or overall capability. Due to the power of the media and the growing low-cost printer segment, 3D printing has become the more recognised term for the industry. Additive manufacturing is now mainly used for more complex processes.

3D printing processes differ from each other in nature of the material used (e.g. plastics, ceramics, metals, resins), technology of deposition, mechanism of formation of the layers or the characteristics of the obtained product (e.g. final shape, surface finish, texture, geometrical shape, mechanical properties). The ASTM classifies these technologies into seven categories of machines based on the additive process involved (Table 2.1) namely material extrusion, material jetting, powder bed fusion, binder jetting, vat photopolymerisation, sheet lamination and directed energy deposition.

3DP is an umbrella term that encompasses a range of different printing technologies with applications across a myriad of fields from medical device manufacturing to aeronautics, robotics, electronics, industrial goods and even the food industry [1]. Within the scope of architecture, 3D printing can be used to justify a proof-of-concept model by manifesting a visual idea to reality almost instantaneously. 3D printing techniques that use metal as a starting material are able to manufacture industrial tools and are capable of repairing and replacing parts of a larger entity. NASA for example have used a 3D printing technology called selective

Table 2.1 Classification of the main 3D printing technologies

ASTM Categories	Technologies	Substrate	Mechanism of layering
Binder jetting	Powder bed inkjet printing S-printing M-printing Theriform™ ZipDose®	Solid particles (plaster, metal, sand, polymer)	A liquid binding agent is selectively deposited to join powder materials
Vat polymerisation	Stereolithography (SLA) Digital light projection (DLP) Continuous layer interface production (CLIP)	Liquid (photopolymer)	Liquid photopolymer in a vat is selectively cured by light-activated polymerisation
Powder bed fusion	Selective laser sintering (SLS) Direct metal laser sintering (DLSM) Selective metal sintering (SLM) Electron beam melting (EBM) Concept laser	Solid particles (metal, plastic, polymer)	Thermal energy selectively fuses regions of a powder bed
Material extrusion	Fused deposition modelling (FDM) Gel/paste extrusion	Filament (thermoplastic polymers e.g. ABS; PLA; PC ULTEM)	Material is selectively dispensed through a nozzle or an orifice
Material jetting	Ink-jet printing Polyjet Thermojet	Liquid (acrylic-based photopolymers, elastomeric photopolymers, wax-like materials)	Droplets of built material are selectively deposited
Directed energy deposition	Electron beam direct Manufacturing Direct metal tooling (DMT) Be additive manufacturing (BeAM)	Wire (metal)	Focused thermal energy is used to fuse materials by melting as they are being deposited
Sheet lamination	Laminated object manufacturing	Sheets	Sheets of material are bonded to form an object

laser sintering (SLS) (discussed further in Chap. 5) to manufacture a rocket engine injector by printing two parts instead of the 115 items required if conventional methods were enforced [1].

More recently, this technology has been applied in medicine to manufacture medical devices [4, 5] and tablets. Thus far, a wide range of simple and complex formulations have been produced including those containing multiple active pharmaceutical ingredients (APIs) [6–8] with different geometries [9, 10] and release characteristics [11–21]. Favourably, this technology enables precise dosages to be deposited based on the initial 'ink' concentration and the physical dimensions of the formulation.

The introduction of 3D printing to pharmaceuticals has the potential to cause a paradigm shift in modern medicine manufacture. Conventionally, formulations are produced using large-scale processes which, whilst being cost-effective for mass manufacture, are inherently time consuming, labour intensive and are limited in terms of dose flexibility. By using 3D printing, the possibility of rapidly dispensing customised dosage forms on-demand could soon be a reality. Indeed this technology can be utilised throughout the drug development timeline, ranging from pre-clinical development and first-in-human (FIH) clinical trials, all the way to front-line medical care. The 3D printer platform could be easily integrated into a vast range of settings such as within clinical trial research facilities or on the front-line in hospitals.

2.1 Industrial Applications of 3D Printing

It has been predicted that in at least five years, 3D printing will govern the manufacturing techniques of industrial tooling, end-use components or perhaps final products. Direct Digital Manufacturing (DDM) is the term given for the fabrication of final products, coined by Stratasys founder and CEO Scott Crump, for the intended use of manufacturers, designers and engineers. This strategy was able to lower manufacturing restrictions in production designs and provided a high degree of customisation. Although it is unlikely for 3D printing to become a complete substitute for traditional manufacturing methods, increasing industrial pioneers have grasped and implemented rapid prototyping techniques.

The early 1990s witnessed an increasing number of commercial operations and subsequently, competing companies in the sector from its disruptive application in manufacturing techniques. This spans from building moulds and small-scale tools rapidly being applied most notably in the aerospace, dental and the automotive sector. The greatest investments of using 3D printing in industry are within the aviation sector. This is due to advantages in fabricating lighter parts and more accurate geometries, often unachievable when using conventional casting and machining tools. Employing a specific 3D technique such as selective laser melting (SLM) (further discussed in Chap. 5) has allowed the 3D printing of metal onto an aircraft component, suggesting its capabilities in servicing technologies and thus, reducing the overall costs of parts if replacement were required. In addition, the automobile manufacturing industry is also starting to explore the potential of 3D printing with the fabrication of a two-passenger hybrid car called the Urbee [1] in 2010. 3D printing has further lent itself to the fashion and retail industry for the customisation of jewellery and even shoes!

In terms of pharmaceutics the first drug product manufactured using 3D printing, Spritam® of Aprecia Pharmaceuticals, was approved by the Food and Drug

Administration (FDA) in 2015. A number of small biotech companies, however, are fast emerging aiming to create personalised dosage forms and medical devices using 3D printing techniques. For example FabRx Ltd., a specialist biotech, is focused on developing 3D printing technologies specifically for fabricating pharmaceuticals.

A timeline of the inventions, development of rapid prototyping techniques and first applications of 3D printing towards engineering and medical routes are outlined in Fig. 2.1.

2.2 3D Printing Software

It is important to acknowledge that 3D printing technologies were not developed as an isolated invention, but rather a result of several innovations from an array of sectors. The development of the rapid prototyping technology in the medical field was only possible due to the advancement of medical imaging techniques, image processing and reconstruction. Similarly, independent of the specific technology used, 3D printing normally involves common steps; 1) the creation of a computer assisted design (CAD) file, 2) its conversion into an .stl file (uploaded to the printer equipment), 3) the printer set-up and finally, 4) the manufacture of the object. Additional steps may include the removal from the build plate and final cleaning. The manufacturing process is considered a single fully computer-controlled procedure, avoiding intermediate stages of production and extensive manual labour. This is one of the major advantages in terms of reduced time and cost of development and production since any desired change in the final object can be achieved by modifying the starting CAD file.

2.3 Binder Jet Printing

Binder jet printing was first invented and patented by Sachs et al. [29] at MIT in 1993. Unlike the majority of 3D printing techniques that manipulate liquids, gels or paste characteristics that result in a solid 3D printed object, binder jetting is a technology that agglomerates powdered material by a layer-by-layer approach, illustrated in Fig. 2.2.

Here, a layer of powder is laid on a building platform (termed a powder bed) which is achieved by raising the base of a powder delivery platform placed adjacent to it. Often employing a blade or a roller, the powder is pushed across the bed. A printer nozzle containing binder liquid is programmed to move across an x-y axis and to spray the liquid onto a thin layer of powder. The liquid binds the powder particles together, causing layer solidification. The fabrication platform is then moved down along the vertical z axis. Subsequently, a thin powder layer is distributed on top and the process is repeated sequentially to fabricate a 3D object [1].

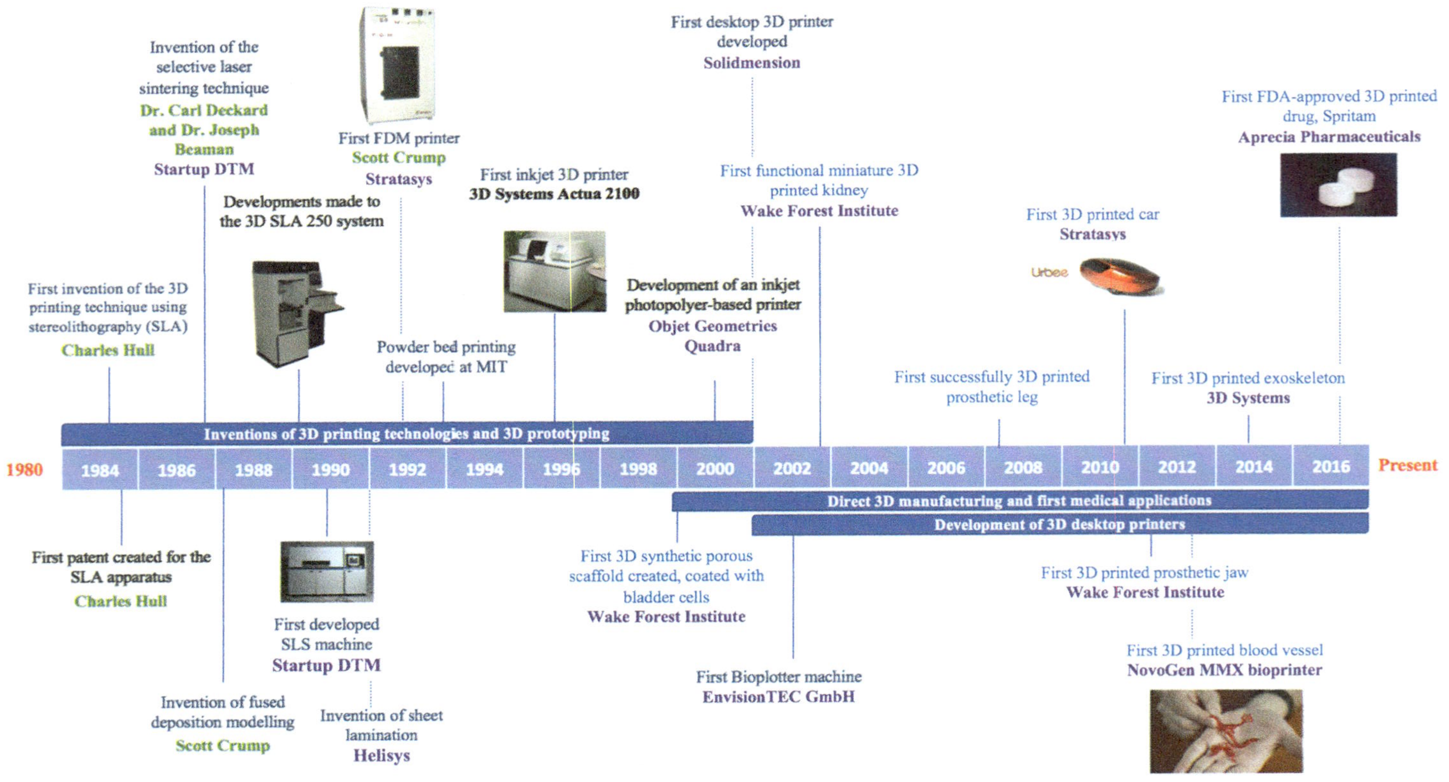

Fig. 2.1 Timeline from 1980–2016: invention of different 3D printing technologies, 3D direct manufacturing and the first medical applications. (Images reprinted with permission from [22–28])

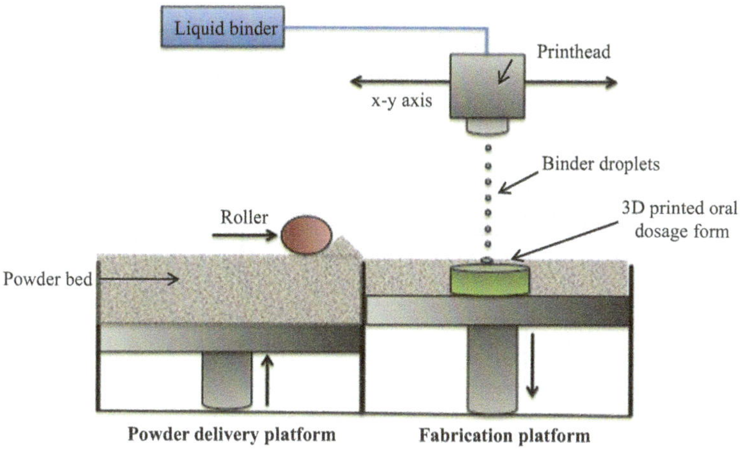

Fig. 2.2 General schematic of the binder jet printing process

Since its invention, binder jet printing has been widely researched in pharmaceutical arena. The first article in this field was published in 1996 which demonstrated the feasibility of using binder jet printing to fabricate a drug delivery device containing synthetic dyes (to act as drug models) [30]. Since then, this technology has been used to create more complex formulations such as zero-order release tablets [31, 32] and erosion-mechanism delayed release tablets [33].

Binder jet printing has been found to be capable of formulating highly porous and fast-dissolving tablets [31]. This benefit was exploited to manufacture Spritam®, the first FDA-approved 3D printed tablet [34] which are oro-dispersible when taken with a sip of water, even with high drug loadings of up to 1000 mg. However, tablets produced with binder jet printing pose a risk of having poor friability and mechanical strength due to the formation of highly porous structures [35]. Furthermore, fabricated objects often require drying steps to evaporate any residual binder or steps to remove excess powder [36]. This is further discussed in Chap. 3.

2.4 *Vat Polymerisation*

Vat polymerisation is a generic term given to a number of 3D printing technologies. A vat or tank of liquid photopolymer is selectively solidified using a laser beam or other light source that is most appropriate for the material chosen. The main vat polymerisation processes include stereolithography (SLA), digital light projection (DLP), daylight polymer printing (DPP), continuous light interface production (CLIP), lithography-based ceramic manufacturing (LCM) and two-photon polymerisation (2PP). Although there are a multitude of technologies surrounding the vat polymerisation process, this book will concentrate on the first 3D printing technology invented and the most applied vat polymerisation technique for pharmaceutical applications, SLA.

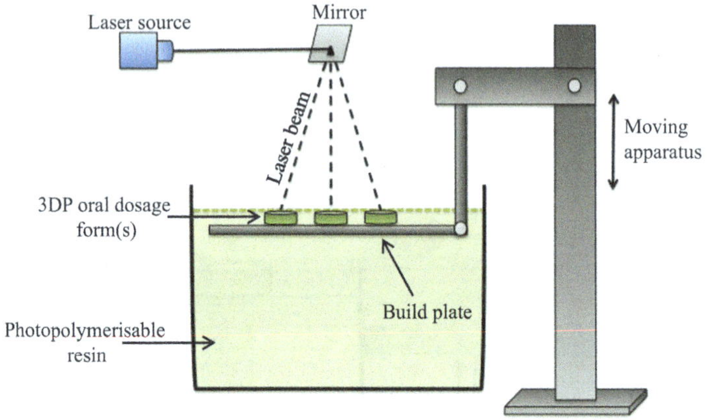

Fig. 2.3 General schematic of the SLA printing process

2.4.1 Stereolithography

The origins of conventional 3D printing can be traced back to the 1980s when the first 3D printing technology was created by Hull. This method involves exposing a photopolymerisable resin to high energy light (e.g. UV light) to induce polymerisation and solidification of the material. Each time, the resin is solidified to a defined depth, the platform is moved down vertically along the z axis, the built layer is recoated with resin and the process is repeated layer-by-layer (Fig. 2.3) [37]. This technology has been widely applied to tissue engineering [38–40] and has more recently been applied to pharmaceutical applications [41–43]. For example, Wang et al. [41] fabricated modified-release tablets containing 4-aminosalicylic acid (4-ASA) and paracetamol using SLA. Compared to alternative 3D printing methods that have typical resolutions between 50 and 200 μm, SLA can produce objects of superior resolution (as low as 20 μm). Furthermore, SLA uses lower localised temperatures (e.g. compared to fused deposition modelling and SLS), being more suitable for thermally labile APIs [37]. However, there are safety concerns with SLA as the commercially available resins and photoinitiators contain functional groups that may exhibit cytotoxicity [44].

2.5 Powder Bed Fusion

Powder based 3D printing processes are characterised by the use of powders, binders, sintering techniques, de-powdering and potential post-processing treatments. Selective laser sintering (SLS), s-printing and m-printing fall under the powder bed fusion umbrella and were all introduced shortly after the invention of the SLA technique. In Chap. 5, we will concentrate on the capabilities and applications of SLS. It is the most commonly used powder bed fusion technique that presents distinct advantages when compared with the latter.

2.5.1 Selective Laser Sintering

Similar to powder bed printing, SLS builds a 3D object from a powder bed. SLS employs a laser, which is directed to draw a specific pattern on the powder bed to cause selective partial or full melting to bind powder particles. Once the layer is sintered, a roller distributes a fresh layer of powder on top of the sintered material [45].

To date, the application of SLS to medical research has been mostly limited to the field of tissue engineering such as the production of scaffolds [46, 47] or drug delivery devices for example [48]. This may be due to the high localised temperatures required to sinter materials which may cause drug or excipient degradation [49]. Because of this, SLS was once regarded as being impossible to print oral medicines. However, in 2017, Fina et al. [50] described the first application of SLS in this field which used an alternative diode laser to manufacture immediate and modified release tablets containing paracetamol at three different drug loadings without drug degradation. More recently, SLS has proven to be amenable for the modern manufacture of printlets with customisable drug release profiles by fabricating complex gyroid lattice structures [51] and orally disintegrating characteristics [52].

Due to the precision of the laser, a higher resolution (~100 μm) can be achieved using SLS compared to other methods (such as FDM and extrusion). Furthermore, SLS is a solvent-free process compared to binder jet printing which may require around 48 h drying time post-fabrication [53]. SLS requires only the preparation of a suitable powder mixture for printing compared to the need for drug-loaded filament production for FDM.

2.6 *Material Extrusion*

This 3D printing technique selectively dispenses a material (either molten/softened or semi-solid in state) through a nozzle or orifice. This printing technology is considered the easiest among customers and it is the technology used in most of the affordable desktop printers. Depending on the materials used, different techniques can be identified such as fused deposition modeling which uses thermoplastic polymers generally in the form of filaments and semi-solid extrusion.

2.6.1 Fused Deposition Modelling

Fused deposition modelling (FDM), also known as fused filament fabrication (FFF), was first commercialised in 1991 [54] and is currently one of the most commonly used, low-cost techniques in 3D printing. The principle of FDM technology is the deposition of thin strands of melted polymer from a filament on a build plate creating one layer of the object to be printed (Fig. 2.4). The build plate then moves down and another layer is deposited. By repeating these steps in a layer-by-layer manner, the final object is obtained [11, 56]. FDM has been the most adopted 3D printing technology to date due to it being inexpensive and easy to use.

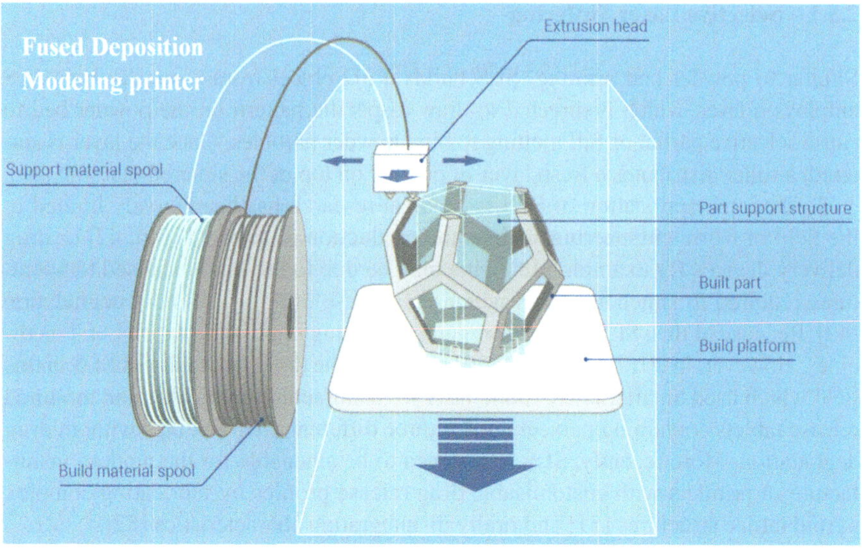

Fig. 2.4 Representation of a FDM printer. (Reprinted with permission from [55])

A tractor wheel arrangement, often implemented in most printers (e.g. MakerBot models), drive polymeric filaments into the heated extrusion head. This procedure also generates the extrusion pressure required to melt or soften the material flow. The extrusion head is governed by a plotting system that conducts the 2D movements. The build plate allows the item to grow in a 3D motion, which is achieved by solidification of the molten extrudate at room temperature. The nozzle orifice determines the overall diameter of the deposited material, however, the vertical descent of the build plate defines the nominal thickness to each layer. From the development of printers, a reverse arrangement can now be followed in which the build plate moves in the horizontal plane (x-y axis) whilst the extrusion head rises vertically.

After the deposition of the last layer, materials that are used to wrap or anchor the fabricated object are removed. Post-processes such as a coating, polishing or sand-papering may be further required to prepare the 3D product for its final form [57].

The starting materials for FDM are often supplied in the form of a filament with a predetermined shape and dimensions that are often produced by hot melt extrusion (HME), further discussed in Chap. 8. Based on the broad demand for plastic products, many filaments are commercially available and are mainly comprised of acrylonitrile butadiene styrene (ABS), polylactic acid (PLA) or high impact polystyrene (HIPS). The application of FDM in other areas, however, requires the development of new filaments of different materials that consider a multitude of properties. This may include its colour, resistance, flexibility, conductivity and biocompatibility.

Thus far, FDM has been the most widely utilised 3D printing technique in pharmaceuticals. A wide variety of formulations have been produced using FDM including immediate release [58] and controlled release tablets [12, 59] as well as orodisperisble films [60]. FDM has a number of advantages over other 3D printing technologies. Firstly, the equipment is cheap with systems ranging from £500 – £2000. Secondly, when considering the filaments used within pharmaceutical research, a variety of biocompatible thermoplastic polymers have been evaluated including polyvinyl alcohol (PVA) [12, 34, 59] and polyacrylates (e.g. Eudragit polymers) [58, 61]. FDM can be used to fabricate many structures quickly and easily with good resolution (~50 – 100 μm). However, the high temperatures used may also cause premature degradation of the API which can be problematic for thermally sensitive drugs. Although a more recent study has shown that FDM is in fact amenable to print drugs with low melting temperatures [62]. This will be discussed in more detail in Chap. 6.

2.6.2 Semisolid Extrusion

Semisolid extrusion is another example of a 3D printing technology under the material extrusion umbrella. Semisolid extrusion employs a syringe-like system that extrudes a paste or gel onto the build plate in a layer-by-layer approach. Similar to FDM, the extruded material then solidifies following cooling or by virtue of solvent evaporation.

Fab@Home and 3D Bioplotter are currently the most common printers that utilise solutions, dispersions, gels and pastes (see Chap. 7). In the case of 3D Bioplotter, which was originally developed to work in sterile conditions for the production of scaffolds, the layers are deposited by one or more syringes under predetermined temperatures and pressures. Fab@Home (Fig. 2.5), however, was the first multi-material 3D printer developed to be commercially available.

A controlled deposition process can be achieved independently and precisely (up to 1 μL). One of the main advantages related to this approach is the ability to apply a large variety of materials, ranging from hydrogels, epoxy resins and even chocolate

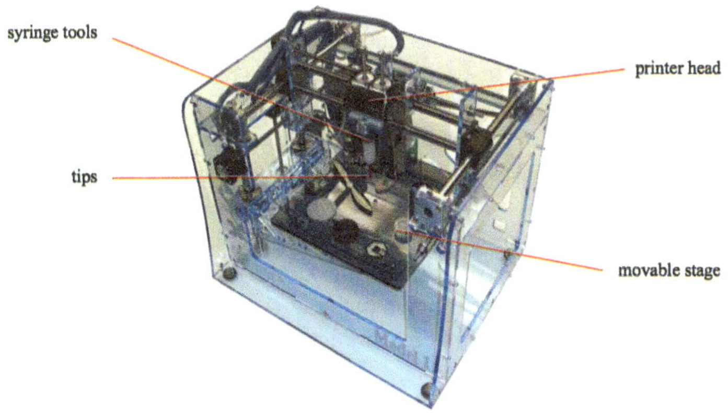

Fig. 2.5 Diagram of a Fab@Home printer. (Reprinted with permission from [63])

and cheese without requiring high temperatures. Semisolid extrusion limitations span from the use of organic solvents incorporated in the solutions, dispersions, gels or pastes that could potentially be hazardous. However, upon successful melting, homogenisation and extrusion, the resulting filament can then be fed into an FDM 3DP system.

In the case of pharmaceuticals, the semi-solids can be formulated by mixing a drug, polymers and solvent(s) in the correct ratio to produce formulations of acceptable viscosity for printing. Khaled et al. [64] has used semi-solid extrusion to produce a number of formulations, including controlled release bi-layer tablets and 'polypills' (a tablet which contains more than one drug). For example, this group used extrusion to formulate three drugs (captopril, glipizide and nifedipine) into the same tablet which permitted drug release via two mechanisms; osmosis and diffusion [6]. Another study demonstrated the ability to print five drugs into the same tablet [7].

2.6.3 Bioprinting

3D printing has more recently been exploited in a multitude of engineering sectors, aerospace and even food. With constant advancements in the 3D printing area, this technology has led to the development of a new scientific area termed as synthetic biology (SynBio). This is the creation of a new variant of 3D printing coined as bioadditive manufacturing or bioprinting [1].

Since the first incidences of printing biologics with cytoscribing governed by Klebe in 1986 [1], 3D bioprinting has achieved significant advances, particularly in the last decade. As conventional techniques are not capable of fabricating objects with the necessary structural, biological and mechanical intricacy, bioprinting offers an approach in overcoming these complexities [65]. 3D bioprinting, like the majority of 3D printing techniques, is an additive manufacturing process that uses bioink to fabricate devices and scaffolds in a layer-by-layer manner (Fig. 2.6).

Bioprinting has been widely adopted in the fabrication of living tissues within the scope of tissue engineering and regenerative medicinal research, transplantation and clinics, and high throughput screening in pharmaceutics. As the bioprinting technique increases in attractiveness and awareness, pharmaceutical and medical fields have developed global institutions with the aim to commercialise and advance our knowledge of bioprinting technologies.

Gaining traction and interest, an example of a recent success in the bioprinting field include the printing of several thyroid glands that were successfully transplanted into live mice by a Russian start-up (Bioprinting Solutions) in November 2015. Following 11 weeks post-surgery, the rats had completely restored their thyroid function. The main objective of Bioprinting Solutions is to 3D print replacement thyroids and complex organs for human patients. An increasing number of research teams around the world have adopted a similar goal, proving that there is a medical demand for tissue engineering.

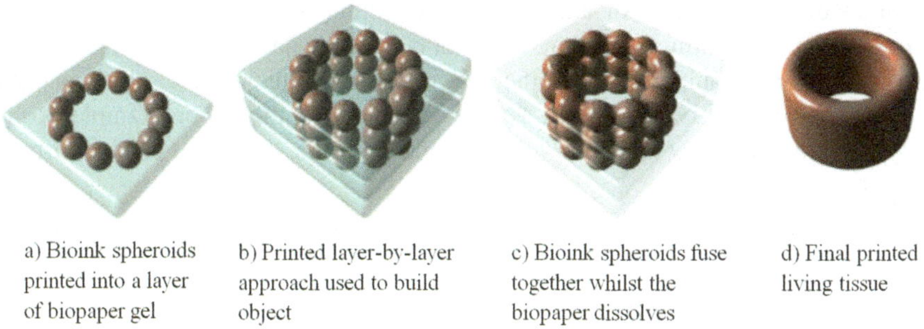

Fig. 2.6 Bioprinting with Bio-ink Spheroids. (Reprinted with permission from [1])

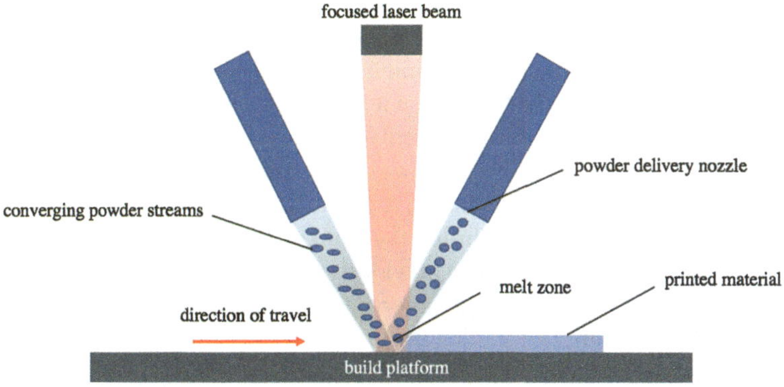

Fig. 2.7 Schematic illustration of the DED printing process

2.7 *Directed Energy Deposition*

Yet another means of creating fused metal objects from powdered material is directed energy deposition (DED), otherwise known as laser powder forming. With this technique, metal powder is directed into a high-power laser beam by a motion-controlled nozzle to deposit as a molten build material (Fig. 2.7). The build platform is often motion-controlled on multiple axis. DED, unlike other 3D printing techniques stated, is not limited to building objects in a successive and flat manner. Multiple companies have adopted and developed the DED process, with the company Optomec Inc. referring to the process as laser engineering net shaping (LENS). TRUMPF Ltd. and BeAM, however, coined the DED process as laser metal deposition (LMD) [1].

Stainless steel, copper, cobalt, nickel, aluminium and titanium can all be employed for DED printing. Unlike with the powder bed fusion technique, the metal powder is fed to the print head and can be manipulated to include different metal compositions during the printout. Fabricated objects of different properties, therefore, can be printed, which is currently unachievable using traditional manufacturing methods. An additional surface-finishing step, however, is often required. As the printing procedure does not occur on a flat bed, DED has a unique advantage when compared with other 3D printing techniques whereby repairing existing objects can occur as well as fabricating new entities. For example, Rolls Royce utilise the DED approach to add metal onto the surface of eroded turbine blades.

2.8 *Sheet Lamination*

Most 3D printing processes either extrude semi-molten material(s), solidify photopolymers or bind powders. Sheet lamination, however, is independent of all these 3D printing strategies as instead of incorporating conventional 3D printing materials (powders or filaments, for example), sheets or paper, plastic or even metal foil adhere together; hence the lamination terminology employed. Laminated object manufacture (LOM), however, is the term given exclusively to the adherence of multiple sheets of paper for the fabrication of a designed product.

With LOM, a sheet of build material (previously supplied with an adhesive backing or coated with adhesive during the build process) is advanced onto a build platform. A laser is then used to cut a previously designed outline into a sheet whilst the build platform descends; the process is continuously repeated until the completion of the design.

Sheet lamination was invented in 1991 by Helisys (Fig. 2.1) which has since gone out of business. However, in 2012, Mcor-Technologies launched three printers that use a patented selective deposition lamination technique to create objects using normal copier paper, thus, is an affordable method of 3D printing when compared with any other technologies.

Another example of a pioneering sheet lamination group is Fabrisonic. This group have developed a technique called ultrasonic additive manufacturing which successfully prints metal tape fused together using high frequency vibrations (20,000 Hz). Similar to the SLM approach, Fabrisonic's technology is capable of welding different metals into the same entity, overcoming challenges that often face traditional manufacturing techniques.

3 3D Printing Implementation in Healthcare

3D printing could be used across a number of healthcare services such as in primary and secondary care in community pharmacies or specialist clinics to larger-scale tertiary care in hospital wards [66].

3.1 Rapid Administration and Improved Medicine Access

On-demand dispensing in a clinical setting could improve medicine access and accelerate discharge times, especially for medicines that require specialist manufacture or extemporaneous preparations. 3D printing, therefore, can rapidly fabricate one-off doses and may achieve tailored doses in time or resource-constrained conditions such as accident and emergency departments, in disaster crises and military operations [67]. In addition, rapid manufacturing allows the fabrication of medicines with poor stability and short shelf lives to be printed on-demand [68] such as certain antibiotics that require a desiccant (nitroglycerin for angina pectoris for example) [69].

3.2 3D Printing and Wearable Diagnostics

3D printing can easily be integrated with other technological advancements of the twenty-first century such as smart health monitors, digital applications and cloud-based computing. It is now feasible for patients to monitor their own vital signs from heart rates and blood pressures using freely accessible applications on smart-devices. Potentially in the near future, clinicians may have access to real-time data within the existing Internet of Things, facilitating an ease of review and modification of treatment or dose. As 3D printers can be easily controlled, a physician or pharmacist could design a 3D printed tablet (herein known as printlets: a term that we have coined to describe 3DP oral dosage forms) and send a prescription to a location with a 3D printer, ready to be fabricated and dispensed [49].

3.3 Challenges to 3D Printing Implementation

As with all new technologies, using 3D printing to produce formulations comes with its own unique challenges that the healthcare industry may have not experienced before. Most notably, there are a number of regulatory and technical challenges that need to be addressed prior to the integration of 3D printed products into standard drug development and clinical practices.

4 Regulatory Requirements

Since the FDA approval of Spritam® and the investment of numerous pharmaceutical companies into the 3D printing of medicines, such as FabRx Ltd. and GlaxoSmithKline, there has been a call from innovators to regulatory agencies to provide guidance and to evaluate how 3D printing can operate within the regulatory framework.

Progress has already commenced in the U.S. following the establishment of the Emerging Technology Team (EET) within the FDA. The EET aims to encourage the application of technological innovation in design and production since the positive shift of 3D printing in manufacturing have been recognised [70]. The Center for Drug Evaluation and Research (CDER) Office of Pharmaceutical Quality has also established a manufacturing science program to investigate the use of innovative technologies in production such as 3D printing and process analytical technologies (PAT) [71]. In 2017, the FDA released a guidance around the additive manufacturing of medical devices and implants [72] which are discussed in subsequent chapters to come. As far as the author is aware of, there are currently no other regulatory guidelines to date that provide information on the 3D printing of other products however.

4.1 Regulatory Considerations of 3D Printing in Pharmaceutics

Regulating the use of 3D printers to produce solid oral dosage forms may be challenging. For example, depending on the location of the printer, whether in a clinical trial site, specials manufacturing facility or hospital ward, the drug product may require classification under each of their specific regulatory pathways. Quality control of the final drug product would also need to be considered. In addition, all aspects of the printing process would need thorough evaluation to ensure consistent product quality from the hardware, raw material suppliers, operator training and quality control. At this point of time, the current commercial 3D printers have not been designed for Good Manufacturing Practices (GMP). Further developments are therefore required to ensure that these unconventional platforms are fit-for-purpose by conforming to such regulations. However, it has been reported that the FDA encourage the development of complex dosage forms and manufacturing processes using science and risk-based approached which may pave the way for 3D printing [67, 73].

5 Conclusion

The 3D printing revolution, like any other, is the product of vision and implementation, brought upon those who were brave enough to make it happen. With the acknowledgement, increasing curiosity and understanding of the capabilities of 3D printing, we stand on the brink of revolution in digital fabrication with the potential to disrupt how future products are made. This advancement is driven by exponential developments in 3D printers themselves and its application in almost all areas of manufacturing and personalisation namely; aeronautics, engineering, architecture and pharmaceutics. 3D printing technologies in their entirety are all deeply discussed in the subsequent chapters to come. The following chapters holistically outline their respective histories, potential advantages and disadvantages and a thorough description of the new and exciting possibilities that can arise, simply by implementing rapid prototyping techniques.

References

1. Barnatt C. 3D printing. In: Explaining the future and create space independent publishing platform. 3rd ed. London; 2016.
2. Markillie P. A third industrial revolution: the economist. 2012. Available from: http://www.economist.com/node/21552901.
3. ASTM. Standard terminology for additive manufacturing technologies. F2792 – 12a. 2012.
4. Huang W, Zheng Q, Sun W, Xu H, Yang X. Levofloxacin implants with predefined microstructure fabricated by three-dimensional printing technique. Int J Pharm. 2007;339(1–2):33–8.
5. Wu W, Zheng Q, Guo X, Sun J, Liu Y. A programmed release multi-drug implant fabricated by three-dimensional printing technology for bone tuberculosis therapy. Biomed Mater. 2009;4(6):065005.
6. Khaled SA, Burley JC, Alexander MR, Yang J, Roberts CJ. 3D printing of tablets containing multiple drugs with defined release profiles. Int J Pharm. 2015;494(2):643–50.
7. Khaled SA, Burley JC, Alexander MR, Yang J, Roberts CJ. 3D printing of five-in-one dose combination polypill with defined immediate and sustained release profiles. J Control Release. 2015;217:308–14.
8. Goyanes A, Wang J, Buanz A, Martinez-Pacheco R, Telford R, Gaisford S, et al. 3D printing of medicines: engineering novel oral devices with unique design and drug release characteristics. Mol Pharm. 2015;12(11):4077–84.
9. Goyanes A, Robles Martinez P, Buanz A, Basit AW, Gaisford S. Effect of geometry on drug release from 3D printed tablets. Int J Pharm. 2015;494(2):657–63.
10. Goyanes A, Scarpa M, Kamlow M, Gaisford S, Basit AW, Orlu M. Patient acceptability of 3D printed medicines. Int J Pharm. 2017;530(1–2):71–8.
11. Goyanes A, Buanz AB, Hatton GB, Gaisford S, Basit AW. 3D printing of modified-release aminosalicylate (4-ASA and 5-ASA) tablets. Eur J Pharm Biopharm. 2015;89:157–62.
12. Goyanes A, Chang H, Sedough D, Hatton GB, Wang J, Buanz A, et al. Fabrication of controlled-release budesonide tablets via desktop (FDM) 3D printing. Int J Pharm. 2015;496(2):414–20.
13. Sadia M, Arafat B, Ahmed W, Forbes RT, Alhnan MA. Channelled tablets: an innovative approach to accelerating drug release from 3D printed tablets. J Control Release. 2018;269:355–63.
14. Okwuosa TC, Pereira BC, Arafat B, Cieszynska M, Isreb A, Alhnan MA. Fabricating a Shell-Core delayed release tablet using dual FDM 3D printing for patient-Centred therapy. Pharm Res. 2017;34(2):427–37.
15. Sadia M, Sośnicka A, Arafat B, Isreb A, Ahmed W, Kelarakis A, Alhnan MA. Adaptation of pharmaceutical excipients to FDM 3D printing for the fabrication of patient-tailored immediate release tablets. Int J Pharm. 2016;513(1–2):659–68.
16. Beck RCR, Chaves PS, Goyanes A, Vukosavljevic B, Buanz A, Windbergs M, Basit AW, Gaisford S. 3D printed tablets loaded with polymeric nanocapsules: An innovtive approach to produce customized drug delivery systems. Int J Pharm. 2017;528(1):268–79.
17. Goyanes A, Fernández-Ferreiro A, Majeed A, Gomez-Lado N, Awad A, Luaces-Rodríguez A, Gaisford S, Aguiar P, Basit AW. PET/CT imaging of 3D printed devices in the gastrointestinal tract of rodents. Int J Pharm. 2018;536(1):158–64.

18. Goyanes A, Fina F, Martorana A, Sedough D, Gaisford S, Basit AW. Development of modified release 3D printed tablets (printlets) with pharmaceutical excipients using additive manufacturing. Int J Pharm. 2017;527(1–2):21–30.
19. Muwaffak Z, Goyanes A, Clark V, Basit AW, Hilton ST, Gaisford S. Patient-specific 3D scanned and 3D printed antimicrobial polycaprolactone wound dressings. Int J Pharm. 2017;527(1–2):161–70.
20. Goyanes A, Kobayashi M, Martínez-Pacheco R, Gaisford S, Basit AW. Fused-filament 3D printing of drug products: microstructure analysis and drug release characteristics of PVA-based caplets. Int J Pharm. 2016;514(1):290–5.
21. Goyanes A, Det-Amornrat U, Wang J, Basit AW, Gaisford S. 3D scanning and 3D printing as innovative technologies for fabricating personalized topical drug delivery systems. J Control Release. 2016;234:41–8.
22. SolidProducts. SLA-2500 product pag: laser innovations; 2012. Available from: http://www.solidimaging.com/2500.htm.
23. 3DPrinterClassified. DTM sinterstation 2000 SLS Machine. 2015. Available from: http://www.3dprinterclassifieds.com/3d-printers/dtm-sinterstation-2000-sls-machine/.
24. Smith S. Evolution of rapid technologies: digital engineering. 2010. Available from: http://www.digitaleng.news/de/evolution-of-rapid-technologies/.
25. RPMI. RPMI equipment: major laboratory equipment: rapid prototyping and manufacturing institute. 2003. Available from: http://rpmi.marc.gatech.edu/about/equip.html.
26. Millsaps BB. Three wheels & two seats: will the 3D printed Urbee 2 be your car of the future?: 3DPrint.com; 2016.Available from: https://3dprint.com/124086/3d-printed-urbee-2-car/.
27. Loubriel A. Scientists can now 3D print working blood vessels: guardian liberty voice. 2014. Available from: http://guardianlv.com/2014/07/scientists-can-now-3d-print-working-blood-vessels/.
28. Aprecia Pharmaceutials. FDA approves the first 3D printed drug product: aprecia pharmaceuticals. 2015. Available from: http://www.multivu.com/players/English/7577251-aprecia-pharmaceuticals-spritam/.
29. Sachs EM, Haggerty J, Cima MJ, Williams PA. inventorsthree-dimensional printing techniques. US5204055.1993.
30. Wu B, Borland SW, Giordano RA, Cima LG, Sachs EM, Cima MJ. Solid free-form fabrication of drug delivery devices. J Control Release. 1996;40:77–87.
31. Yu DG, Shen XX, Branford-White C, Zhu LM, White K, Yang XL. Novel oral fast-disintegrating drug delivery devices with predefined inner structure fabricated by three-dimensional printing. J Pharm Pharmacol. 2009;61(3):323–9.
32. Wang CC, Tejwani Motwani MR, Roach WJ, Kay JL, Yoo J, Surprenant HL, et al. Development of near zero-order release dosage forms using three-dimensional printing (3-DP) technology. Drug Dev Ind Pharm. 2006;32(3):367–76.
33. Katstra WE, Palazzolo RD, Rowe CW, Giritlioglu B, Teung P, Cima MJ. Oral dosage forms fabricated by three dimensional printing. J Control Release. 2000;66:1–9.
34. Pharmaceutials A. Manufactured using 3D printing. 2015. Available from: http://www.spritam.com/-/hcp/zipdose-technology/manufactured-using-3d-printing.
35. Yu DG, Zhu LM, Branford-White CJ, Yang XL. Three-dimensional printing in pharmaceutics: promises and problems. J Pharm Sci. 2008;97(9):3666–90.
36. Yu DG, Branford-White C, Yang YC, Zhu LM, Welbeck EW, Yang XL. A novel fast disintegrating tablet fabricated by three-dimensional printing. Drug Dev Ind Pharm. 2009;35(12):1530–6.
37. Melchels FP, Feijen J, Grijpma DW. A review on stereolithography and its applications in biomedical engineering. Biomaterials. 2010;31(24):6121–30.
38. Melchels FP, Feijen J, Grijpma DW. A poly(D,L-lactide) resin for the preparation of tissue engineering scaffolds by stereolithography. Biomaterials. 2009;30(23–24):3801–9.
39. Elomaa L, Teixeira S, Hakala R, Korhonen H, Grijpma DW, Seppala JV. Preparation of poly(epsilon-caprolactone)-based tissue engineering scaffolds by stereolithography. Acta Biomater. 2011;7(11):3850–6.

40. Scalera F, Esposito Corcione C, Montagna F, Sannino A, Maffezzoli A. Development and characterization of UV curable epoxy/hydroxyapatite suspensions for stereolithography applied to bone tissue engineering. Ceram Int. 2014;40(10):15455–62.
41. Wang J, Goyanes A, Gaisford S, Basit AW. Stereolithographic (SLA) 3D printing of oral modified-release dosage forms. Int J Pharm. 2016;503(1–2):207–12.
42. Martinez PR, Goyanes A, Basit AW, Gaisford S. Fabrication of drug-loaded hydrogels with stereolithographic 3D printing. Int J Pharm. 2017;532(1):313–7.
43. Martinez PR, Goyanes A, Basit AW, Gaisford S. Influence of geometry on the drug release profiles of stereolithographic (SLA) 3D printed tablets. AAPS PharmSciTech. 2018; https://doi.org/10.1208/s12249-018-1075-3.
44. Arcaute K, Mann BK, Wicker RB. Stereolithography of three-dimensional bioactive poly(ethylene glycol) constructs with encapsulated cells. Ann Biomed Eng. 2006;34(9):1429–41.
45. Lee J-Y, An J, Chua CK. Fundamentals and applications of 3D printing for novel materials. Appl Mater Today. 2017;7:120–33.
46. Duan B, Wang M, Zhou WY, Cheung WL, Li ZY, Lu WW. Three-dimensional nanocomposite scaffolds fabricated via selective laser sintering for bone tissue engineering. Acta Biomater. 2010;6(12):4495–505.
47. Du Y, Liu H, Yang Q, Wang S, Wang J, Ma J, et al. Selective laser sintering scaffold with hierarchical architecture and gradient composition for osteochondral repair in rabbits. Biomaterials. 2017;137:37–48.
48. Leong KF, Chua CK, Gui WS, Verani. Building porous biopolymeric microstructures for controlled drug delivery devices using selective laser sintering. Int J Adv Manuf Technol. 2006;31(5–6):483–9.
49. Alhnan MA, Okwuosa TC, Sadia M, Wan KW, Ahmed W, Arafat B. Emergence of 3D printed dosage forms: opportunities and challenges. Pharm Res. 2016;33(8):1817–32.
50. Fina F, Goyanes A, Gaisford S, Basit AW. Selective laser sintering (SLS) 3D printing of medicines. Int J Pharm. 2017;529(1–2):285–93.
51. Fina F, Goyanes A, Madla CM, Awad A, Trenfield SJ, Kuek JM, Patel P, Gaisford S, Basit AW. 3D printing of drug-loaded gyroid lattices using selective laser sintering. Int J Pharm. 2018; https://doi.org/10.1016/j.ijpharm.2018.05.044.
52. Fina F, Madla CM, Goyanes A, Zhang J, Gaisford S, Basit AW. Fabricating 3D printed orally disintegrating printlets using selective laser sintering. Int J Pharm. 2018;541(1–2):101–7.
53. Rowe CW, Katstra WE, Palazzolo RD, Giritlioglu B, Teung P, Cima MJ. Multimechanism oral dosage fors fabricated by three dimensional printing. J Control Release. 2000;66:11–7.
54. Crump SS. Apparatus and method for creating three-dimensional objects US 5121329 A. 1992.
55. Additively. Fused Deposition Modeling (FDM): additively.com; 2015. Available from: https://www.additively.com/en/learn-about/fuseddeposition- modeling.
56. Goyanes A, Buanz AB, Basit AW, Gaisford S. Fused-filament 3D printing (3DP) for fabrication of tablets. Int J Pharm. 2014;476(1–2):88–92.
57. Awad A, Trenfield SJ, Gaisford S, Basit AW. 3D printed medicines: A new branch of digital healthcare. Int J Pharm. 2018;548(1):586–96.
58. Pietrzak K, Isreb A, Alhnan MA. A flexible-dose dispenser for immediate and extended release 3D printed tablets. Eur J Pharm Biopharm. 2015;96:380–7.
59. Gioumouxouzis CI, Katsamenis OL, Bouropoulos N, Fatouros DG. 3D printed oral solid dosage forms containing hydrochlorothiazide for controlled drug delivery. J Drug Deliv Sci Technol. 2017;40:164–71.
60. Jamróz W, Kurek M, Łyszczarz E, Szafraniec J, Knapik-Kowalczuk J, Syrek K, et al. 3D printed orodispersible films with aripiprazole. Int J Pharm. 2017;533(2):413–20.
61. Melocchi A, Parietti F, Maroni A, Foppoli A, Gazzaniga A, Zema L. Hot-melt extruded filaments based on pharmaceutical grade polymers for 3D printing by fused deposition modeling. Int J Pharm. 2016;509(1–2):255–63.

62. Kollamaram G, Croker DM, Walker GM, Goyanes A, Basit AW, Gaisford S. Low temperature fused deposition modeling (FDM) 3D printing ofthermolabile drugs. Int J Pharm. 2018;545(1–2):144–52.
63. Lipson H. Fab@Home model 1 3D printer: Wikipedia. 2016. Available from: https://en.wikipedia.org/wiki/File:Fab@Home_Model_1_3D_printer.jpg.
64. Khaled SA, Burley JC, Alexander MR, Roberts CJ. Desktop 3D printing of controlled release pharmaceutical bilayer tablets. Int J Pharm. 2014;461(1):105–11.
65. Ji S, Guvendiren M. Recent advances in bioink design for 3D bioprinting of tissues and organs. Front Bioeng Biotechnol. 2017;5:23.
66. Trenfield SJ, Awad A, Goyanes A, Gaisford S, Basit AW. 3D printing pharmaceuticals: drug development to front-line care. Trends Pharmacol Sci. 2018;39(5):440–51.
67. Awad A, Trenfield SJ, Goyanes A, Gaisford S, Basit AW. Reshaping drug development using 3D printing. Drug Discov Today. 2018; https://doi.org/10.1016/j.drudis.2018.05.025.
68. Alomari M, Mohamed FH, Basit AW, Gaisford S. Personalised dosing: printing a dose of one's own medicine. Int J Pharm. 2015;494(2):568–77.
69. Kommanaboyina B, Rhodes CT. Trends in stability testing, with emphasis on stability during distribution and storage. Drug Dev Ind Pharm. 1999;25(7):857–68.
70. FDA. Emerging technology program maryland, US: food and drug administration. 2017. Available from: https://www.fda.gov/AboutFDA/CentersOffices/OfficeofMedicalProductsandTobacco/CDER/ucm523228.htm.
71. Markarian J. FDA and the emerging technology of 3D printing. Pharm Technol. 2016;40(8). Available from: http://www.pharmtech.com/fdaand- emerging-technology-3d-printing.
72. FDA. Technical considerations for additive manufactured devices: food and drug administration. 2017. Available from: https://www.fda.gov/downloads/MedicalDevices/DeviceRegulationandGuidance/GuidanceDocuments/UCM499809.pdf.
73. Yu LX, Akseli I, Allen B, Amidon G, Bizjak GT, Boam A, Caulk M, Doleski D, Famulare J, Fisher AC, Furness S, Hasselbalch B, Havel H, Hoag SW, Iser R, Johnson BD, Ju R, Katz P, Lacana E, Lee SL, Lostritto R, McNally G, Mehta M, Mohan G, Nasr M, Nosal R, Oates M, O'Connor T, Potti J, Raju GK, Ramanadham M, Randazzo G, Rosencrance S, Schwendeman A, Selen A, Seo P, Shah V, Sood R, Thien MP, Tong T, Trout BL, Tyner K, Vaithiyalingam S, VanTrieste M, Walsh F, Wesdyk R, Woodcock J, Wu G, Wu L, Yu L, Zezza D. Advancing product quality: a summary of the second FDA/PQRI conference. AAPS J. 2016;18(2):528–43.

Chapter 3
Binder Jet Printing in Pharmaceutical Manufacturing

Sarah J. Trenfield, Christine M. Madla, Abdul W. Basit, and Simon Gaisford

Abstract Binder jet printing is arguably the most successful three-dimensional printing (3DP) technology in the pharmaceutical industry to date. In 2015, the binder jet process was adapted as an alternative mass manufacturing technique to enable the production of Spritam® (the first 3D printed tablet) approved by the Food and Drug Administration (FDA). Binder jet printing is expected to continue making a widespread impact to formulation manufacture over the next decade. In particular, binder jet printing offers benefits of producing oral dosage forms with unique release characteristics ranging from fast-dissolving through to controlled-release platforms. This chapter aims to discuss the history and methodology of binder jet printing, pharmaceutical and medical applications, considerations for formulations development and advantages and disadvantages of such processes in the pharmaceutical space.

Keywords 3D printing · Additive manufacturing · Printing medicines · Personalized medications · Drug delivery systems · Digital healthcare

1 Introduction

Binder jet printing was first invented and patented by Sachs et al. [1] (US Patent USOO5340656A) at the Massachusetts Institute of Technology (MIT) in 1993. Unlike the majority of 3D printing techniques that manipulate liquids, gels or paste characteristics that result in a solid 3D printed object, binder jet printing is a

S. J. Trenfield (✉) · C. M. Madla
Department of Pharmaceutics, UCL School of Pharmacy, University College London, London, UK
e-mail: sarah.trenfield.16@ucl.ac.uk

A. W. Basit · S. Gaisford
Department of Pharmaceutics, UCL School of Pharmacy, University College London, London, UK

FabRx Ltd., Ashford, Kent, UK

A. W. Basit, S. Gaisford (eds.), *3D Printing of Pharmaceuticals*, AAPS Advances in the Pharmaceutical Sciences Series 31,
https://doi.org/10.1007/978-3-319-90755-0_3

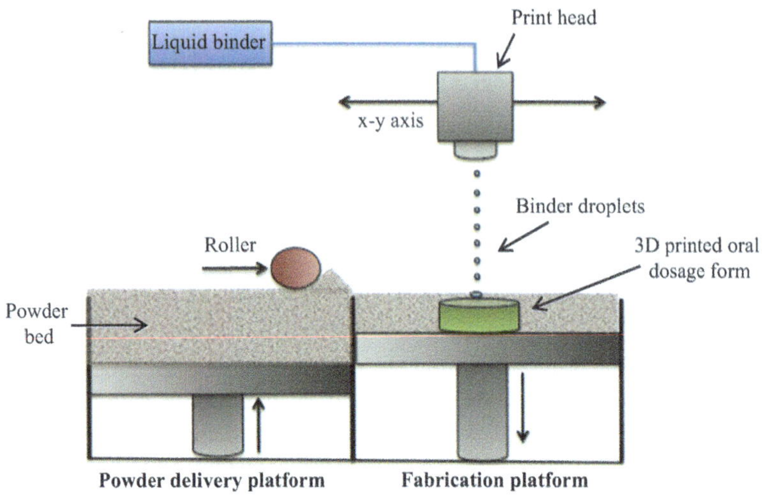

Fig. 3.1 General schematic of the binder jet printing process

technology that fuses powdered material by a layer-by-layer approach illustrated in Fig. 3.1. This technology is arguably the most successful to date for pharmaceutical applications; in 2015, a modified binder jet printing process was used to manufacture the first FDA-approved 3D printed tablet, Spritam® by Aprecia Pharmaceuticals [2]. As such, there has been a growing interest in using this technique to manufacture oral dosage forms. This chapter will discuss the methodology of binder jet printing, pharmaceutical applications to date, process and formulation considerations and the advantages and disadvantages of binder jet printing as an ameanable 3D printing technique for pharmaceuticals.

2 Binder Jet Methodology

The requirements for object design using binder jet printing are similar to the other 3D printing methods; a digital file is created using computer aided design (CAD) software or based on an imaging technique, such as magnetic resonance imaging (MRI), to create a 3D representation of the object. In particular, using CAD, the digital file is saved as an .stl format which is readable by the printer software. In the case of pharmaceuticals, this gives the significant advantage of enabling objects to be tailored to the patient or study requirements by controling the shape, size, dosage and other formulation characteristics.

The binder jet printing process is based on conventional ink-jet printing technology and involves five main steps (Fig. 3.2). Step 1: A printer nozzle containing binder (and/or drug) liquid is programmed to move across an x-y axis

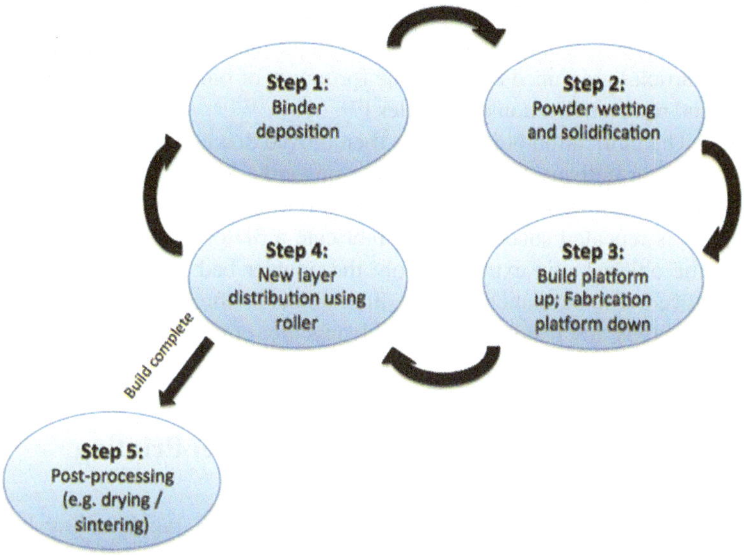

Fig. 3.2 Five stages of the binder jet printing process

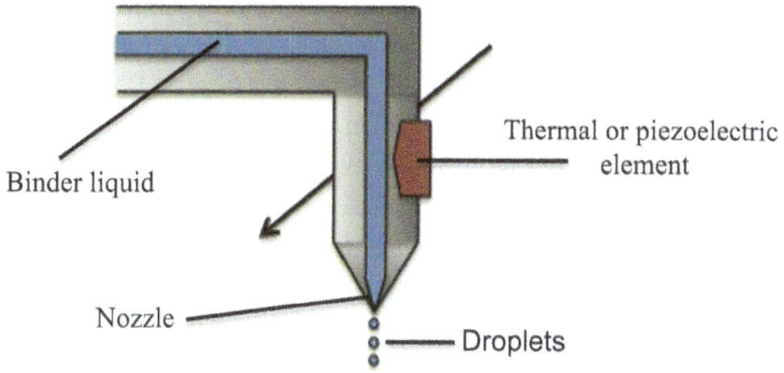

Fig. 3.3 General schematic of the DoD inkjet printing process

and to spray the liquid onto a loose powder bed; Step 2: The powder particles are wetted by the droplet which results in local hardening and layer solidification.

To control droplet deposition, binder jet printers commonly use two different kinds of drop-on-demand (DoD) heads; piezoelectric and thermal heads (Fig. 3.3) [3]. The thermal system is comprised of a print head containing a small liquid reservoir that contains the binder liquid. A current is pulsed through a resistive element in the print head, causing an internal temperature rise and subsequent vaporisation, nucleation and expansion of a bubble which imparts sufficient energy to eject a droplet [4, 5].

In the piezoelectric system a high voltage pulse is applied to a piezoelectric material element (usually lead zirconate titanate) which induces a volumetric change in the liquid. When the voltage is applied to the element, it deforms, imparting sufficient pressure build-up resulting in ejection of the fluid [6]. Following droplet deposition, the method of powder solidification is similar to the mechanisms for wet granulation whereby particles are joined by either the formation of binder bridges or by the dissolution and re-crystallisation of particles [7].

Step 3: The fabrication build plate is then moved down along the vertical z axis, with the powder delivery platform moving upwards; Step 4: Subsequently, a thin powder layer is distributed using a roller on top of the previously bound layer and the process is repeated successively to fabricate a 3D object; And finally, Step 5: involves the object being extracted from the powder bed with the excess powder being removed. The permanence of the final object is often achieved using thermal sintering whereby any residual volatile solvents are removed [8, 9].

3 Pharmaceutical Applications of Binder Jet Printing

Early research in the 1990s used binder jet printing for rapid tooling by using powdered metals and ceramics as this technology was initially developed to specifically cater for this application [10]. The flexibility that binder jet printing offers is unprecedented. From a materials perspective, virtually any powder can be used to produce parts, providing that it is combined with a suitable binder. Moreover, the binder jet process can be scaled up to create large objects (metre range) as well as small objects (few millimetres) [11].

This technology offers excellent precision enabling the creation of objects with highly complex geometries that would otherwise be difficult or impossible to create with conventional manufacturing processes. Due to the capability for complex structure fabrication, there has been a growing interest in using this technique in medicine for the creation of scaffolds particularly within bone tissue engineering. For example, several papers have reported creating scaffolds composed of the biomaterial hydroxyapatite [12–14], as well as polylactic acid (PLA) combined with a bioactive calcium phosphate [15].

In recent years, the application of binder jet printing to pharmaceuticals has been increasingly researched [16]. Following the patenting of the binder jet process in 1993, the applications of binder jetting in healthcare (e.g. production of medical devices and oral dosage forms) were licensed exclusively to Theric Ltd., which was trademarked as the TheriForm™ process in 1996 [17]. Between 1999 and 2000, two different methodologies of using binder jetting to formulate pharmaceuticals were developed [16];

1. The print drug approach: drug-loaded liquids (with or without binder) are deposited onto an excipient powder bed and
2. The print glue approach: binder liquids are deposited onto a drug-loaded powder bed.

The first paper relating to the use of binder jetting in the pharmaceutical field was published in 1996 which demonstrated the feasibility of fabricating a sustained

Table 3.1 Examples in the literature of formulations produced using binder jet printing

Formulation type	Powder bed	Binder and solvent	Drug compound	Drug incorporation	References
Sustained release tabular device	PEO	PCL and chloroform	Methylene blue Alizarin yellow	Ink	[18]
Rapidly dispersible tablets	PVP K30 Mannitol Lactose	PVP K30, ethanol and water	Paracetamol	Powder bed	[19]
Cubic dosage forms with near zero order release	Kollidon SR HPMC	PVP K17 PVP K17 + TEC	Pseudoephedrine	Ink	[20]
Erosion mechanism delayed release tablets	Eudragit® E100	Ethanol	Chlorphenamine	Ink	[21]
	Eudragit® RLPO	Acetone			
	Lactose	PVP + water	Fluoroscein	Ink	
Delayed release tablets	Ethylcellulose HPMC E50 PVP K30	Ethylcellulose + ethanol	Paracetamol	Powder bed	[22]
Rapidly dispersible tablets (TheriFlash™)	Mannitol Maltitol Maltodextrin PVP 25	PVP 25 + aqueous buffer	Captopril	Ink	[23]
Zero order release tablets	Ethylcellulose HPMC E50 PVP K30	Ethylcellulose	Paracetamol	Powder bed	[24]

release drug delivery device via the print drug approach [18]. The formulation was comprised of the bioresorbable polymers polycaprolactone (PCL) and polyethylene oxide (PEO) containing synthetic dyes to act as drug models. This was considered a significant milestone in the history of binder jet printing for pharmaceuticals. However, the solvents dichloromethane and chloroform were used in this work that, due to their toxicity and difficulty in ensuring the removal of trace quantities, were deemed unsuitable for medical use. Since then, binder jetting has been explored as a technique to create a wide range of different tablets, ranging from controlled-release formulations to fast dissolving tablets (Table 3.1). These different types of formulations will be discussed in turn.

3.1 *Controlled-Release Formulations*

Thus far, binder jet printing has been used to create a range of complex formulations that would not easily be produced by conventional manufacturing technologies. For example, Kastra et al. [21] successfully developed two types of controlled-release

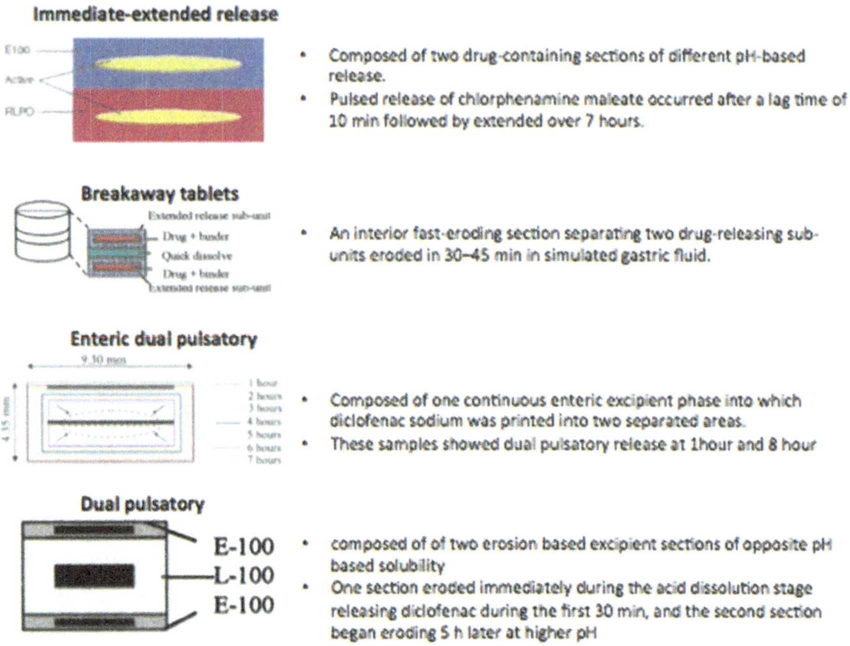

Fig. 3.4 Schematic and characteristics of four different dosage forms produced using binder jet printing. (Adapted with permission from source: [27])

tablets using binder jetting. Erosion-mechanism delayed release tablets were constructed by depositing Eudragit® E-100 binder (which dissolves at < pH 5.0 [25]) at the external walls of the tablet, with a 30% w/w aqueous solution of chlorphenamine maleate at the inner region. Their research group also created diffusion-mediated tablets by depositing Eudragit® RLPO (a water insoluble yet permeable polymer regardless of pH [26]) at the external walls of the tablet and a centre core impregnated with chlorphenamine maleate. In both cases, binder droplet placement was shown to be highly accurate, producing tablets with excellent content uniformity containing as low as microgram levels of drug.

Following on from this, Rowe et al. [27] fabricated four different types of complex oral dosage forms (immediate-extended release tablets, breakaway tablets, enteric dual pulsatory tablets and dual pulsatory tablets) using binder jetting. Each formulation was composed of different excipients and hence, demonstrated differing drug release profiles. A detailed explanation of these dosage forms is shown in Fig. 3.4.

Binder jet printing has been used to develop other complex formulations such as near zero-order controlled-release formulations. For the past two decades, pharmaceutical research in the scope of drug delivery has identified that zero-order release kinetics is a favourable profile for the majority of bioactive agents [28]. However, many of the strategies of achieving zero-order release are either too

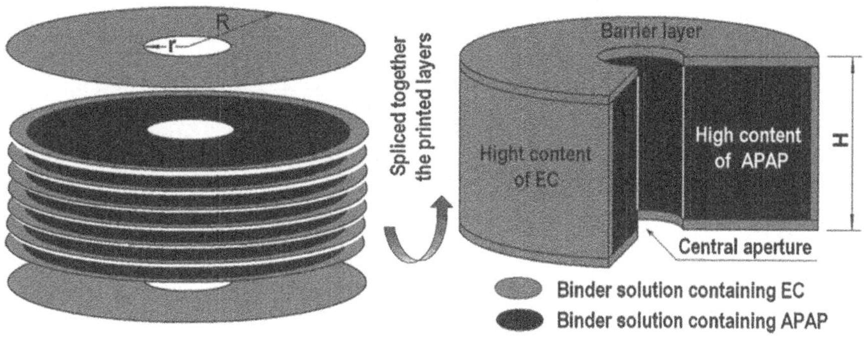

Fig. 3.5 Schematic representation of a layer-by-layer doughnut structure as a novel dosage form. (Reprinted with permission from source: [22])

difficult to manufacture or require the use of overly sensitive drug carrier systems. Different types of tablet geometries have been trialled and tested to achieve zero-order drug release profiles ranging from a doughnut shape [29], rectangular [30], biconvex [31], parabolic [32] and core-in-cup approaches [28]. Considering all these profiles, the doughnut-shaped tablet designed with one or more cylindrical holes provide constant drug release and have gained the most attention [33].

More recent work has been conducted to replicate a doughnut-shaped dosage form. However, methods of achieving this geometry have been challenging with attempts to bore a hole in the centre of a tablet by virtue of high-speed drilling emerging. From the advancements and recognition of binder jet printing in the pharmaceutical field, Yu et al. [22] demonstrated that a doughnut-shaped, multi-levelled structure could be printed that was capable of combining varying drug concentrations and release profiles into one dosage form, simply by altering prototyping parameters (Fig. 3.5).

This study printed a variety of different drug delivery devices containing paracetamol of varying diameters, heights and concentrations of drug and excipients. Favourably, the doughnut-shaped tablets were found to perform superior to standard paracetamol tablets, achieving linear release profiles. The 3D printed devices were further able to delay drug release by controlling the concentration of ethylcellulose, a commonly used controlled-release polymer. Paracetamol was released through an erosion process from the surface and the inner orifices of the doughnut-shaped device. This resulted in a consistent delayed-release profile for the whole drug dissolution assessment. The overall dose, therefore, with a 3D printed doughnut-shaped device, could be adjusted independently by changing the height of the dosage form.

Another research group demonstrated the production of a drug delivery device with a near zero order release profile. Wang et al. [20] used binder jet printing to formulate pseudoephedrine hydrochloride with various polymers to control drug release and to act as drug carriers. The drug device composed of an immediate release core and an outer shell that regulated the drug release rate.

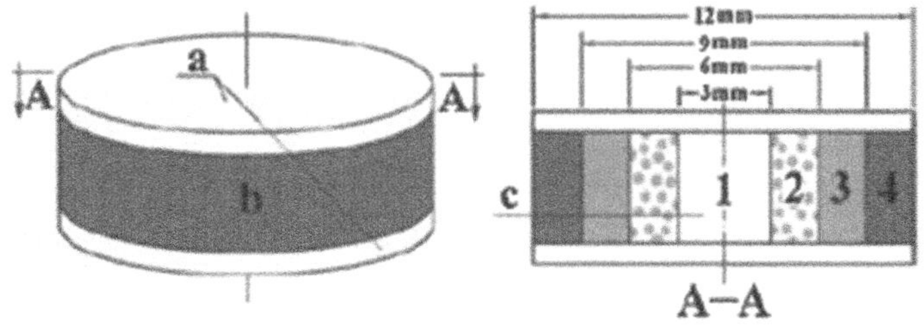

Fig. 3.6 Schematic of tablets with material gradients fabricated by binder jet printing: (*a*) release-retarding layer, (*b*) drug containing region, (*c*) gradients of release-retarding materials. (Reprinted with permission from source: [24])

The study demonstrated the successful development of a near zero-order controlled release formulation, modulated by the ratios of polymers used in the drug design. The core-shell structure allowed minimal erosion to occur, thus, drug release was achieved via diffusion through the polymer matrix. Furthermore, the dosage form was insensitive to changes in pH of the dissolution medium and the paddle-stirring rate. Favourably, results showed that there were linear correlations between *in vivo* absorption profiles to *in vitro* dissolution parameters indicating a good relationship for *in vivo*/*in vitro* performance.

Through a different principle, Yu et al. [24] also fabricated a drug delivery system with zero-order release characteristics (Fig. 3.6). Their research group used the print glue approach and incorporated paracetamol, hydroxypropylmethylcellulose (HPMC), polyvvinylpyrollidone (PVP) K30 and paracetamol in the powder bed. A release-retarding binder (ethylcellulose in ethanol) was deposited at the top and bottom of the tablet, with the concentration of release-modulated materials being gradually reduced from the periphery to the centre of the tablet, whilst the void space was increased. The gradient distributions were achieved by depositing the ethylcellulose ethanolic binder a pre-defined amount of times at each region. For example, the first print pass involved the binder being deposited at regions 1, 2, 3 and 4; the second print pass at regions 2, 3 and 4; the third print pass at regions 3 and 4 and the final print pass at region 4 only (Fig. 3.6).

In vitro dissolution studies evaluated that the drug was released by surface erosion and that 98% of the drug (68% of the tablets' weight) was released linearly in 12 h. By using the print glue approach, higher drug loadings were achieved; in this study, the core was able to contain ~70%v/v, compared to a drug in ink approach which only reached ~20%v/v [17, 20]. Desired release profiles are often accompanied with complex design features and complicated multi-step manufacturing processes. This study demonstrated that the 3D printing of release-retardation materials can prepare tablets with a distinctly high dosage and special design features to manipulate desired drug release properties, corroborating with the research conducted by Yu et al. and [22] and Wang et al. [20].

3.2 Rapidly Dissolving Tablets

Advantageously, binder jet printing has been found to be capable of formulating highly porous and fast-dissolving tablets [34]. This is due to the process depositing binder onto a loose powder bed and hence relying solely on the 'glue' to maintain object integrity (instead of mechanical compression force). As such, rapidly dispersible tablets can be formulated due to the ease of water penetration throughout the formulation matrix. For example, Lee et al. [23] developed fast disintegrating tablets containing captopril, termed TheriFlash™ dosage forms, which were composed of a placebo shell and active core. The authors evaluated 13 different formulations and found that factors such as excipient selection, level of excipients and binder saturation all contributed to the TheriForm™ process. A rapidly dispersing dosage form with adequate properties was developed using mannitol (97.5%), maltitol (2.5%) and a binder saturation level of 45%.

In 2009, Yu et al. [19] developed a rapidly disintegrating drug delivery device containing paracetamol. To formulate this, the outer regions of the tablet were bound by selectively depositing binder liquid, with the core containing loose, unbound powder (Fig. 3.7). This dosage form demonstrated acceptable pharmacotechnical properties and acceptable hardness (63.4 N/cm^2), however failed friability measurements (3.55% mass loss). Average disintegration times (23.4 s) and wetting times (67.6 s) were rapid, and dissolution tests showed that 98.5% of the drug was released in under 2 min. As such, this paper showed the capability for the production of rapidly dispersing 3D printed tablets by creating a unique inner structure.

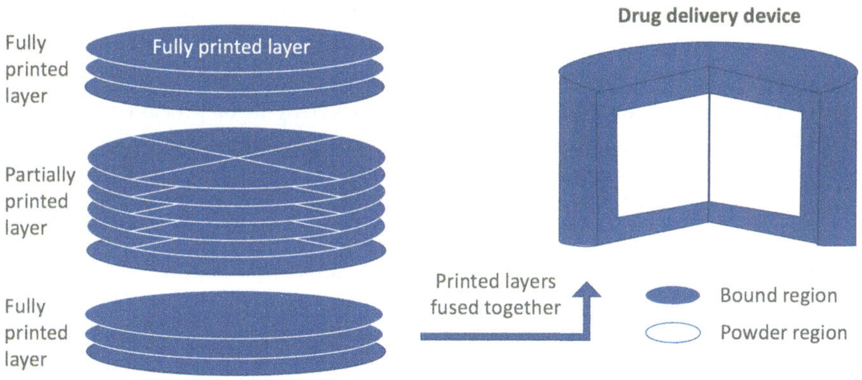

Fig. 3.7 Schematic of the fast-dissolving drug delivery device

3.3 Spritam®: The First FDA Approved 3D printed Product

Levetiracetam is a second-generation anti-epileptic drug that is used as an adjunctive treatment (taken in addition to other epilepsy medication) in patients with partial-onset seizures [35]. Commercially available levetiracetam formulations include oral solutions, extended-release tablets and as an injectable. Poor adherence to therapy, however, is a prevalent issue for paediatric and geriatric patients, and for those who experience difficulties in swallowing, demonstrated in almost one third of psychiatric patients. Although the oral route remains to be the preferred route of drug administration, for this specific group of patients, improvements in dosing, ease of use and drug tolerability are required for overall patient satisfaction. Orally disintegrating tablets (ODTs) were developed in the early 1990s to attempt to alleviate swallowing difficulties of conventional tablets and capsules. However, ODT products are often limited to 10–30 mg of drug loadings. To overcome the low dose capabilities of ODT products, binder jet printing has been employed to fabricate highly porous and rapidly disintegrating oral dosage forms, capable of incorporating up to 1000 mg of levetiracetam.

Unlike other epilepsy medications (or any other drug on the market, in fact), Spritam® of Aprecia Pharmaceuticals is the first FDA-approved 3D printed prescription drug [36, 37]. By virtue of its binder jet printing manufacturing process which forms highly porous tablets, Spritam® is capable of rapidly dissolving in the mouth with an average disintegration time of 11 s (ranging from 2 to 27 s), providing the intake of a small sip of liquid. The liquid formulation that binds levetiracetam and excipients to fabricate the drug contains taste-masking technologies, thus improving patient compliance.

Aprecia Pharmaceuticals adapted the binder jet process to create a 'conveyer belt-like' process, enabling the mass manufacture of oral dosage forms in a single unit operation (ZipDose® technology) [2]. In particular, the ZipDose® technology is based on the foundation of binder jet printing whereby the powder blend is deposited as a thin, single layer. The binding fluid is then distributed onto the surface which allows adherence of the powder particles. The layer-by-layer process is repeated until the finished product is achieved. This manufacturing process is the first and only patented drug-formulation platform that uses 3D printing to enable the final dosage form to be available in doses of up to 1000 mg.

When comparing Spritam® to a conventional tablet formulation of levetiracetam, no clinically significant effects on vital signs or on the oral mucosa were noted. A single dose of the 3D printed 1000 mg tablet was safe and well tolerated by subjects participating in a bioequivalence study [36]. The study concluded that the 3D printed ODT was equivalent in the rate of absorption and bioavailability to the immediate-release levetiracetam tablet.

4 Benefits and Drawbacks

For solid oral dosage form production, binder jet printing provides a number of benefits. One of the main advantages presented by binder jet printing is its ability to manufacture low bulk density, highly porous, fast dissolving tablets with high drug loadings [21], a benefit exploited to fabricate the epilepsy medication, Spritam®. In addition, based on the reasonable selection of excipients, some, or even all, of the active ingredients may be present in the amorphous phase, which could be particularly useful for drugs with poor solubility [19]. However, the high porosity of the tablet (and hence disintegration time) must be balanced against the mechanical strength of the formulation. Due to the absence of mechanical compression, the only force keeping the particles together is the binder. As such, formulations may have poor mechanical strength and poor friability which may not be suitable for handling or packaging processes [37]. During the formulation development phase, selection of suitable powdered excipients, binder and printing parameters will be required to balance these two factors.

Favourably, compared to other 3D printing methods, binder jet printing can be used to create highly intricate and complex formulations. This is because the powder bed can act as a physical support, negating the need for supports or rafts, opening up a wider range of possibilities for tablet designs and structures that can be produced. However, binder jet printing does not currently support the production of hollow formulations as the powder is distributed layer by layer [38]. The spatial control of binder deposition can be accurately controlled however, enabling the production of highly complex formulations with different release profiles. In addition, taste masking technologies can already be incorporated in the binding formulation with binder jet printing, negating the need for lengthy coating processes [37]. Furthermore, the printing process itself can be conducted at room temperature, negating the issue of *in situ* drug degradation that could occur with other 3D printing methods.

In the highly competitive market that is the pharmaceutical industry, the timescale for the developmental pipeline is crucial. Binder jet printing offers a faster and more economical solution for manufacturing, capable of fabricating individual or small batches of formulations, extending even up to the industrial scale, exemplified by the ZipDose® technology platform. For example, previous research has shown that five different formulations were successfully manufactured synchronously using the same powder blend due to the manipulation of binding agent volume deposition [39]. Exploiting this advantage could print a multitude of drug product prototypes, significantly reducing the development time of new products when compared with conventional methods.

However, although binder jet printing holds much promise and excitement for manufacturing and pharmaceutical applications, this technology is certainly still in its developmental stages [40, 41]. Particular challenges for binder jet printing to overcome (considering the pharmaceutical industry alone) include process/manufacturing optimisation techniques and the selection of appropriate excipients. In particular, the operator is required to have a detailed understanding of the materials and processes to ensure that the dosage form is produced accurately, robustly and reproducibly. For example, binder liquid characteristics (e.g. viscosity, surface tension and saturation) are the most critical parameters that affect the behaviour of ink droplets and binder jetting [42]. The ideal binder should exhibit a sufficiently low viscosity as the internal force produced by the print heads is often fairly low. However, this needs to be balanced against a high surface tension to prevent premature ink ejection from the nozzle [43]. Furthermore, the binder must be capable of drying or curing rapidly so that the next layer of powder can be applied [10]. Drug or binder saturation also needs to be considered; in the case of drugs, exceeding the saturation point can cause 'salting out' or crystallisation of the drug causing nozzle blockage, affecting the printing quality [44].

Furthermore, some of the pharmaceutical binders typically used may need to be dissolved or suspended in solvents such as ethanol, chloroform or dichloromethane which are not ideal for formulation manufacture for two reasons. Firstly, some of these solvents can be inherently toxic, warranting them unsuitable for medical use. Secondly, this typically requires 24–48 h of post-processing drying time to remove any leftover solvents and to secure consolidation of the powder particles [19], as well as removal of loose powder particles.

Other considerations include the characteristics of the powder (e.g. flowability, wettability and reactivity). To ensure effective building of the 3D object, consideration of powder flowability is essential. This parameter is influenced by particle shape and size with high powder flowability enabling the roller to distribute thin homogenous layers of powder to produce objects of high resolution. Typically, a high resolution object can be produced by using powder with a small particle size [42]. However, a balance must be struck between resolution and flowability as finer powders pose the risk of agglomeration, negatively impacting powder flow [45]. Poor flowability can prevent excess powder from being removed in the last step of the fabrication process. To aid in powder flow, a lubricant (such as magnesium stearate) could be added to the process to reduce the interparticulate interactions. Other parameters such as powder wettability and reactivity with the binder need to be considered during the formulation development stage.

5 Conclusion

The discovery of binder jetting in 1993 was considered a significant milestone in 3D printing history and has since played an increasingly critical role in a number of fields, including pharmaceuticals. Importantly, this technology was used to produce Spritam®, the first FDA-approved 3D printed tablet and is expected to continue making a widespread impact to formulation manufacture over the next decade. Progress is already being made with cutting-edge research being published every day detailing the new possibilities that binder jet printing can bring. This chapter has discussed the history and methodology of binder jetting, pharmaceutical and medical applications, considerations for formulations development and advantages and disadvantages of such processes in the pharmaceutical space.

References

1. Sachs EM, Haggerty J, Cima MJ, Williams PA. InventorsThree-dimensional Printing Techniques. US5204055.1993.
2. Aprecia Pharmaceuticals. 3D printing - ZipDose technology. 2015. Available from: https://aprecia.com/zipdose-platform/3d-printing.php.
3. Day SP, Shufflebottom L. Evidential value from inkjet printers. Probl Forensic Sci. 2001;XLVI:356–74.
4. Buanz AB, Saunders MH, Basit AW, Gaisford S. Preparation of personalized-dose salbutamol sulphate oral films with thermal ink-jet printing. Pharm Res. 2011;28(10):2386–92.
5. Vuddanda PR, Alomari M, Dodoo CC, Trenfield SJ, Velga S, Basit AW, Gaisford S. Personalisation of warfarin therapy using thermal ink-jet printing. Eur J Pharm Sci. 2018;117:80–7.
6. Noguera R, Lejeune M, Chartier T. 3D fine scale ceramic components formed by ink-jet prototyping process. J Eur Ceram Soc. 2005;25(12):2055–9.
7. Norman J, Madurawe RD, Moore CM, Khan MA, Khairuzzaman A. A new chapter in pharmaceutical manufacturing: 3D-printed drug products. Adv Drug Deliv Rev. 2017;108:39–50.
8. Daly R, Harrington TS, Martin GD, Hutchings IM. Inkjet printing for pharmaceutics – a review of research and manufacturing. Int J Pharm. 2015;494(2):554–67.
9. Alomari M, Mohamed FH, Basit AW, Gaisford S. Personalised dosing: printing a dose of one's own medicine. Int J Pharm. 2015;494(2):568–77.
10. Sachs E, Cima M, Cornie J. Three-dimensional printing: rapid tooling and prototypes directly from a CAD model. CIRP Ann. 1990;39(1):201–4.
11. Lu K, Reynolds WT. 3DP process for fine mesh structure printing. Powder Technol. 2008;187(1):11–8.
12. Cox SC, Thornby JA, Gibbons GJ, Williams MA, Mallick KK. 3D printing of porous hydroxyapatite scaffolds intended for use in bone tissue engineering applications. Mater Sci Eng C. 2015;47(Supplement C):237–47.
13. Duan B, Wang M, Zhou WY, Cheung WL, Li ZY, Lu WW. Three-dimensional nanocomposite scaffolds fabricated via selective laser sintering for bone tissue engineering. Acta Biomater. 2010;6(12):4495–505.
14. Leukers B, Gülkan H, Irsen SH, Milz S, Tille C, Schieker M, et al. Hydroxyapatite scaffolds for bone tissue engineering made by 3D printing. J Mater Sci Mater Med. 2005;16(12):1121–4.
15. Serra T, Planell JA, Navarro M. High-resolution PLA-based composite scaffolds via 3-D printing technology. Acta Biomater. 2013;9(3):5521–30.
16. Hsiao W-K, Lorber B, Reitsamer H, Khinast J. 3D printing of oral drugs: a new reality or hype? Expert Opin Drug Deliv. 2017;15(1):1–4.
17. Awad A, Trenfield SJ, Gaisford S, Basit AW. 3D printed medicines: A new branch of digital healthcare. Int J Pharm. 2018;548(1):586–96.
18. Wu BM, Borland SW, Giordano RA, Cima LG, Sachs EM, Cima MJ. Solid free-form fabrication of drug delivery devices. J Control Release. 1996;40(1):77–87.
19. Yu D-G, Branford-White C, Yang Y-C, Zhu L-M, Welbeck EW, Yang X-L. A novel fast disintegrating tablet fabricated by three-dimensional printing. Drug Dev Ind Pharm. 2009;35(12):1530–6.
20. Wang CC, Tejwani Motwani MR, Roach WJ, Kay JL, Yoo J, Surprenant HL, et al. Development of near zero-order release dosage forms using three-dimensional printing (3-DP) technology. Drug Dev Ind Pharm. 2006;32(3):367–76.
21. Katstra WE, Palazzolo RD, Rowe CW, Giritlioglu B, Teung P, Cima MJ. Oral dosage forms fabricated by three dimensional printing™. J Control Release. 2000;66(1):1–9.
22. Yu DG, Branford-White C, Ma ZH, Zhu LM, Li XY, Yang XL. Novel drug delivery devices for providing linear release profiles fabricated by 3DP. Int J Pharm. 2009;370(1–2):160–6.

23. Lee K-J, Kang A, Delfino JJ, West TG, Chetty D, Monkhouse DC, et al. Evaluation of critical formulation factors in the development of a rapidly dispersing captopril oral dosage form. Drug Dev Ind Pharm. 2003;29(9):967–79.
24. Yu DG, Yang XL, Huang WD, Liu J, Wang YG, Xu H. Tablets with material gradients fabricated by three-dimensional printing. J Pharm Sci. 2007;96(9):2446–56.
25. Jung J-Y, Yoo SD, Lee S-H, Kim K-H, Yoon D-S, Lee K-H. Enhanced solubility and dissolution rate of itraconazole by a solid dispersion technique. Int J Pharm. 1999;187(2):209–18.
26. Pawar PK, Gautam C. Design, optimization and evaluation of mesalamine matrix tablet for colon drug delivery system. J Pharm Investig. 2016;46(1):67–78.
27. Rowe CW, Katstra WE, Palazzolo RD, Giritlioglu B, Teung P, Cima MJ. Multimechanism oral dosage forms fabricated by three dimensional printing™. J Control Release. 2000;66(1):11–7.
28. Danckwerts MP, Watt JGVD, Moodley I. Zero-order release of theophylline from a Core-in-cup tablet in sequenced simulated gastric and intestinal fluid. Drug Dev Ind Pharm. 1998;24(2):163–7.
29. Sundy E, Danckwerts MP. A novel compression-coated doughnut-shaped tablet design for zero-order sustained release. Eur J Pharm Sci. 2004;22(5):477–85.
30. Mishra DS, Yalkowsky SH. A flat circular hole device for zero-order release of drugs: characterization of the moving dissolution boundary. Pharm Res. 1990;7(11):1195–7.
31. Narasimhan B, Langer R. Zero-order release of micro- and macromolecules from polymeric devices: the role of the burst effect. J Control Release. 1997;47(1):13–20.
32. Bayomi MA. Geometric approach for zero-order release of drugs dispersed in an inert matrix. Pharm Res. 1994;11(6):914–6.
33. Cleave JP. Some geometrical considerations concerning the design of tablets. J Pharm Pharmacol. 1965;17(11):698–702.
34. Yu DG, Shen XX, Branford-White C, Zhu LM, White K, Yang XL. Novel oral fast-disintegrating drug delivery devices with predefined inner structure fabricated by three-dimensional printing. J Pharm Pharmacol. 2009;61(3):323–9.
35. FDA. Keppra XR (Levetiracetam) extended-release tablets. 2009. Available from: https://www.accessdata.fda.gov/drugsatfda_docs/nda/2008/022285s000_TOC.cfm.
36. Boudriau S, Hanzel C, Massicotte J, Sayegh L, Wang J, Lefebvre M. Randomized comparative bioavailability of a novel three-dimensional printed fast-melt formulation of Levetiracetam following the Administration of a single 1000-mg dose to healthy human volunteers under fasting and fed conditions. Drugs R&D. 2016;16(2):229–38.
37. Yu DG, Zhu L-M, Branford-White CJ, Yang XL. Three-dimensional printing in pharmaceutics: promises and problems. J Pharm Sci. 2008;97(9):3666–90.
38. Kalaskar D. 3D printing in medicine. 1st ed. Duxford: Woodhead Publishing; 2017.
39. Monkhouse D, Kumar S, Rowe C, Yoo J. Rapid prototyping and manufacturing process. 2003. Available from: http://www.google.com/patents/US20030173695.
40. Trenfield SJ, Awad A, Goyanes A, Gaisford S, Basit AW. 3D printing pharmaceuticals: drug development to frontline care. Trends Pharmacol Sci. 2018;39(5):440–51.
41. Awad A, Trenfield SJ, Goyanes A, Gaisford S, Basit AW. Reshaping drug development using 3D printing. Drug Discov Today. 2018; https://doi.org/10.1016/j.drudis.2018.05.025.
42. Butscher A, Bohner M, Hofmann S, Gauckler L, Müller R. Structural and material approaches to bone tissue engineering in powder-based three-dimensional printing. Acta Biomater. 2011;7(3):907–20.
43. Vaezi M, Yang S. 2 – Freeform fabrication of nanobiomaterials using 3D printing. Rapid prototyping of biomaterials: Woodhead Publishing; 2014. p. 16–74.
44. Raijada D, Genina N, Fors D, Wisaeus E, Peltonen J, Rantanen J, et al. A step toward development of printable dosage forms for poorly soluble drugs. J Pharm Sci. 2013;102(10):3694–704.
45. Krantz M, Zhang H, Zhu J. Characterization of powder flow: static and dynamic testing. Powder Technol. 2009;194(3):239–45.

Chapter 4
The History, Developments and Opportunities of Stereolithography

Pamela Robles Martinez, Abdul W. Basit, and Simon Gaisford

Abstract Stereolithography (SLA) is an additive manufacturing technique that uses light as the source of energy. SLA 3D printing (3DP) was the first rapid prototyping method developed and perhaps the most popular due to its superior resolution and accuracy. Due to its versatility, SLA has been widely studied for its use in tissue engineering or in dentistry. In the pharmacoprinting field, SLA offers a great potential for fabricating complex drug delivery systems as well as approaching the need to manufacture personalised medicine. Despite this, research in the use of SLA 3DP in the pharmaceutical area is still limited. This chapter presents an overview of the fundamental science behind the photopolymerisation process and the SLA 3DP technologies available. A variety of its biomedical uses are presented. The multiple potential pharmaceutical applications and recent advances are reviewed, along with the advantages and limitations of this rapid prototyping technique for the manufacture of modern medicines.

Keywords Three-dimensional printing · Medical devices · Bioprinting · Personalized medications · Pharmacoprinting · Drug delivery systems

P. Robles Martinez (✉)
Department of Pharmaceutics, UCL School of Pharmacy,
University College London, London, UK
e-mail: pamela.martinez.13@ucl.ac.uk

A. W. Basit · S. Gaisford
Department of Pharmaceutics, UCL School of Pharmacy,
University College London, London, UK

FabRx Ltd., Ashford, Kent, UK

A. W. Basit, S. Gaisford (eds.), *3D Printing of Pharmaceuticals*,
AAPS Advances in the Pharmaceutical Sciences Series 31,
https://doi.org/10.1007/978-3-319-90755-0_4

1 Introduction

Three-dimensional printing (3DP) is a rapid prototyping (RP) technique first patented in 1986 by Charles Hull who is known as the father of 3DP [1]. Rapid prototyping can be defined as the fabrication of physical objects based on electronic data from a computer-aided design (CAD) file [2]. There are different types of RP techniques classified on the state of the initial material used for fabrication [3] namely;

- Bulk solid-based RP including interference melting and laminated object manufacturing
- Liquid-based RP including melting and solidification (e.g. fused deposition modelling), laser (e.g. stereolithography and thermal polymerization) and
- Powder-based RP including 1 component (e.g. selective laser sintering) or a combination of a component and a binder (e.g. 3D printing)

This chapter is dedicated to stereolithography (SLA), its history and general uses. Basic concepts will firstly be introduced followed by its applications in pharmaceutics.

SLA is perhaps the most popular RP technique as it offers various advantages over other 3D printing techniques [4] including higher accuracy and resolution (as low as 1 μm for microSLA and 10 μm for SLA – versus 100 μm for FDM) [5] and the ability to fabricate objects with complex internal geometries. For processing polymeric materials for instance for thermoplastic parts, technologies based on extrusion or melting (for example fused deposition modelling or selective laser sintering) are commonly used. However, they are considered as "slow" methods with a lower resolution and a weak layer adhesion compared to light-based technologies [6]. When thermal degradation is a concern, SLA offers an advantageous alternative when compared with technologies such as FDM [7, 8].

Printing resolution, surface smoothness, processing speed and bonding between layers are highly improved by the outstanding spatial control and flexibility offered by light-based 3D printing methods [6, 9].

Despite its numerous advantages, SLA also has its limitations that will be later discussed. The main disadvantage is the rather limited set of photopolymerisable materials, their relatively high cost and perhaps the need of post-fabrication steps for improving mechanical properties and removing unreacted material. An additional possible inconvenience is the cost of desktop SLA and DLP 3D printers which are still more expensive when compared to other 3D printing technologies (like FDM). Although, the latest 3D printers are now available for less than $5000 with resolution of as low as 10 μm [10].

1.1 History of stereolithography

It has been more than 30 years since the first 3D printer and first 3D printed product (Fig. 4.1) was manufactured. The first 3D printer consisted of a stereolithographic apparatus which could produce objects by consecutively solidifying layers of a photocurable material one on top of the other. This rapid-solidification process was triggered by UV light as source of energy [1].

Hull was working with a company that used UV light to apply thin layers of plastic veneers on furniture, paper products and table tops. During this time he developed the idea of placing multiple layers of this materials, one on top of each other, to build 3D objects in almost any conceivable shape [13].

Hull developed a system where a photopolymer solution contained in a vat was solidified following exposure to a light source which traced the shape of the object. This tracing process was repeated several times, placing solidified layers one on top of each other until the object was finalized [13].

From the first 3D printed object, this manufacturing technique has rapidly grown to become what is expected to be the next industrial revolution [14]. With exciting new applications emerging, the possibilities today seem to be endless.

Multiple sectors are now exploring the potential of 3D printing to boost their production and explore new applications, crossing previous boundaries in fabrication methods. For instance, researchers at Virginia Tech recently published a paper about mask-projection stereolithography 3D printing with a high temperature stable material (Kapton, temperature at approximately 600°C) which could not previously be processed with any other traditional techniques due to its molecular composition. This material provides now promising applications for the aerospace, automotive and microelectronics fields [15].

The applications of 3DP, again, are widespread among a multitude of industries (Fig. 4.2) including the facilitating of prototype production and even within the scope of fashion with 3D printed clothes [16], shoes [17] and jewellery [12]. Other

Fig. 4.1 The first 3D printed part: an eye-wash cup fabricated by stereolithography. (Reprinted with permission from [11])

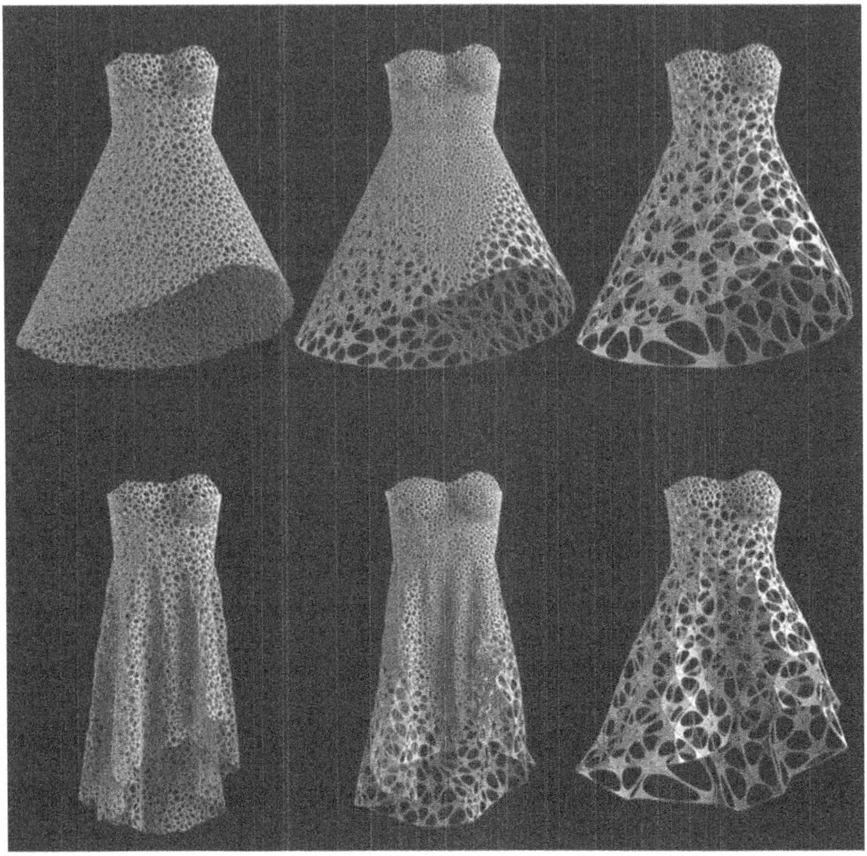

Fig. 4.2 3D printed Kinematics dress from product design and manufacturing company, Nervous System. (Reprinted with permission from [12])

exciting applications include the recently accomplished 3D printed bridge railings in China [18], 3D printed building structures at an architectural scale [19] and even satellites that are 3D printed and launched into space [20].

The biomedical area in particular has greatly benefited from the unique perks of 3D printing. For example, Fig. 4.3 shows results from researchers who recently developed a bioink containing cardiac progenitor cells for fabricating a stem cell patch showing to be a potential suitable platform to deliver cells with a high retention capability as well as to be able to regenerate ischemic tissue [21]. Another group [22] recently used human embryonic kidney cells and bovine mesenchymal stem cells to 3D print constructs that were viable and able to differentiate into cartilage-like structures secreting type-II collagen.

3D printing is thus, a highly flexible manufacturing technique, facilitating the production of complex objects whose potential is being recognised amongst researchers and innovators across multiple areas.

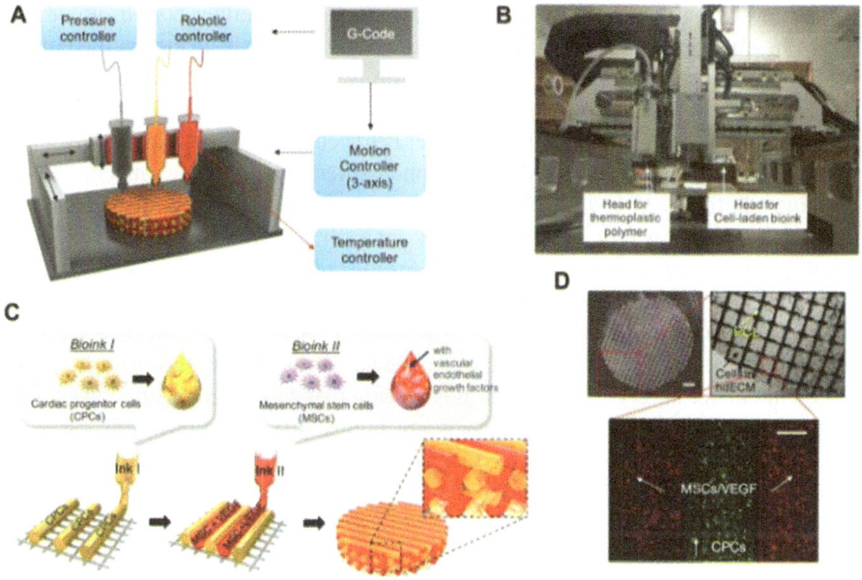

Fig. 4.3 3D printing using bioinks, left for fabricating a stem cell patch. (Reprinted with permission from [21])

2 Stereolithography

SLA is defined as a process for three-dimensional production in which an object is built from a CAD file by the obtained by the consequent deposition of thin layers of a liquid resin that is hardened by its exposure to a scanning laser beam [23].

For their use in 3DP, CAD files need to be transformed into an RP format. This is done by a tessellation process where an algorithm creates a representation of the surface of the object covered by triangles. This triangular mesh is stored as an .stl file, hence, the .stl file is a group of oriented triangles that enable the description of 3D models and it is the usual format for data input for RP systems [24]. The .stl file contains information about the surface of the object excluding texture and colour etc. [25]. This is then exported to a printer software where the electronic object is then sliced in the x-y plane and built in a layer-wise manner in the z-direction [4].

2.1 The Stereolithographic 3DP Process

The basic mechanism of SLA 3DP consists of the computer-spatially-controlled solidification of a liquid resin by a process called photopolymerisation.

Depending on the set up of the 3D printer, they can be either bottom-up or top-down. In a bottom-up set up, the light source (being a laser) is located at the bottom of the printer and the moving platform on top, whereas in a top-down set up, the laser is at the top and the platform at the bottom.

During the printing process the first layer is traced by the laser (directed by galvo mirrors) in the x- and y-axes and the first photocured layer becomes attached to the building platform. After this, the building platform moves up on the z-axis to a defined distance depending on the thickness of each layer (z-resolution). The liquid resin in the container is then redistributed for the solidification of the next layer on top of the previous one. This process is repeated until the object is complete, however, the base and supports (which are usually required) are still needed to be removed. The final object is washed to remove any uncured liquid resin and depending on the material, UV exposure for post-curing may be required [9].

In a top-down system, the process is not highly different, but in this case, the platform is submerged into a VAT containing the photocurable resin. The pattern is then traced by the laser and then the platform moves up one layer thickness. After this, a re-coater blade replaces the photopolymerisable liquid to repeat the hardening process for the rest of the layers one at a time. In a DLP 3D printer, instead of a focused laser tracing the pattern, the source of light is a digital projector which shines the whole image for a complete layer all at once, making the process faster than with a laser beam.

Figures 4.4, 4.5, and 4.6 show a general schematic representation of the bottom-up, top-down and DLP 3D printers. DLP is a faster method since the layers are projected and hence, photocured all at once. In SLA types of 3D printers, the speed is determined by how fast the laser draws the entire layer however, high laser speeds can reduce the quality of the printed object.

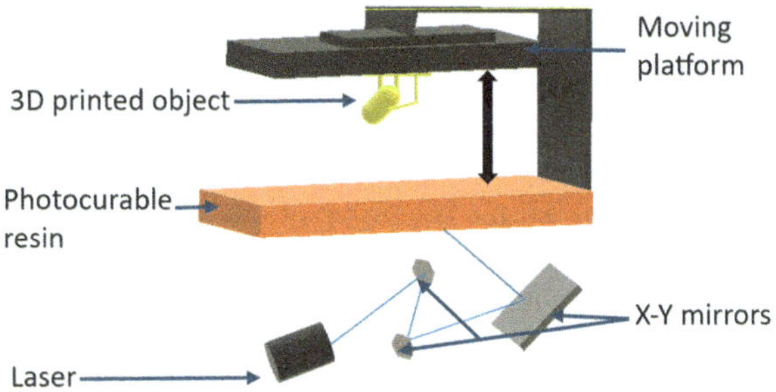

Fig. 4.4 Bottom up SLA 3D printer

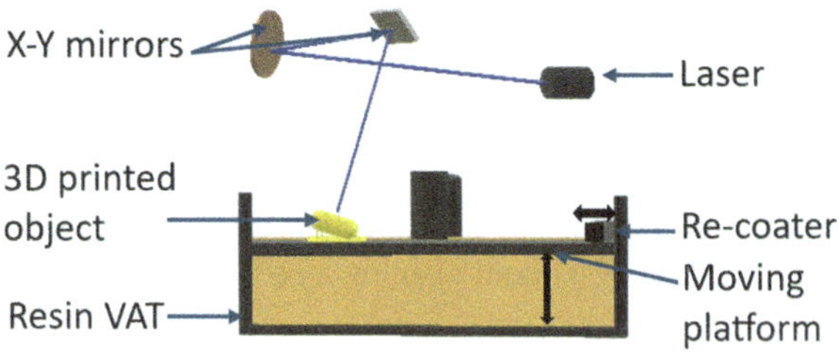

Fig. 4.5 Top-down SLA 3D printer

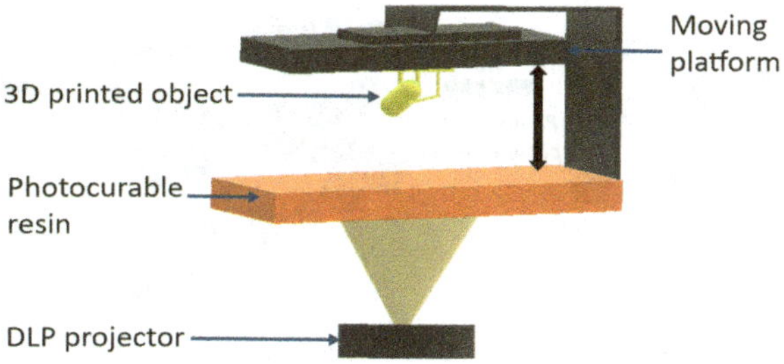

Fig. 4.6 DLP 3D printer

In terms of z-resolution (layer thickness), SLA can achieve < 10 μm whilst DLP can fabricate from 30 to 100 μm which are superior to any other types of 3D printers [5]. A higher resolution 3DP technique (from 1 to 10 μm) known as microstereolithography is capable of hardening each slice of material with a focused laser beam (a few micrometres in diameter). This allows the fabrication of structures to be more precise than those produced by conventional stereolithographic systems [26].

A new faster method was recently developed. Continuous liquid interface production (CLIP) takes advantage of the inhibition that oxygen can create in a photocuring process. By creating a "dead zone" between an oxygen-permeable build window and the surface of the object being fabricated. CLIP is able to produce parts in a non-stop manner (not stepwise like traditional SLA methods including light exposure, resin renewal and part-movement as separate stages). However, the speed-limiting factors in CLIP are the resin cure rate and resin viscosity. Since it is not a stepwise layer production, objects are finished with a smoother surface [27].

2.2 *Photopolymerisation*

The term polymerisation refers to a process where a polymer is synthesised in a chain reaction. This reaction can be triggered by different stimuli including temperature, light, x-rays and γ-rays, an electron beam, plasma, microwaves or pressure [28]. When the reaction is initiated following exposure to light of a suitable wavelength, the process is known as photopolymerisation and the resulting solid is usually insoluble [29].

Photopolymerisation reactions are considered to be a green technology requiring low electrical power input, low temperature of operation and being solvent-free [28]. The photopolymerisation process requires at least three components: a source of light, a photopolymerisable monomer/oligomer and a photoinitiator (PI). Additional components may be incorporated to modify the resin properties such as flow, wetting, dispersion agents, inhibitors, fillers, plasticizers, matting or gloss agents, pigments and light stabilizers as required) [29].

The process can be divided in three general stages [30] (Fig. 4.7);

Initiation: in this first step of the polymerisation process, a molecule known as photoinitiator undergoes a reaction upon exposure to light producing initiating

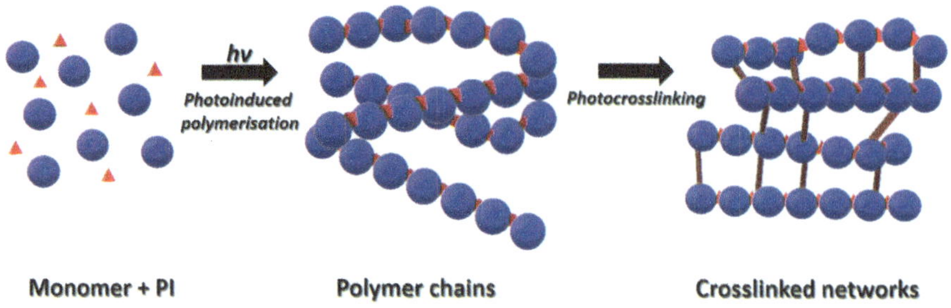

Fig. 4.7 Schematic representation of a photopolymerisation process

species, such as free radicals. These reactive species are able to attack monomer units and consequently add other monomers/oligomers.

Propagation: the reactive species propagate between the monomers/oligomers and the functional groups in these react with each other to start polymerisation and, since it is a 3-dimensional process, crosslinking occurs.

Termination: the polymerisation process is finished when covalent bonds are formed between the crosslinked networks. Since the last stage of the process is photocrosslink, the photocuring process can too, be divided into two stages: (1) photo-induced polymerisation and (2) photocrosslinking. The former refers to a chain reaction where addition of monomers in a linear manner occurs whilst the latter refers to a process where a crosslink is formed between two macromolecular chains [30].

Figure 4.8 shows a general mechanism for a free radical photopolymerisation process. Briefly, as mentioned, exposure to light induces the production of excited states of the PI which yields to the production of initiating species (radical, catonic or anionic depending on the nature of the PI) which are then able to attack monomer units and start the polymerisation process.

2.2.1 Light Sources

Monochromatic light is defined as an electromagnetic wave that travels at a speed c (3×10^8 m/s in vacuum) at a specific wavelength (λ, usually given in nm). Hence it has a frequency (v) of $v = c/\lambda$, usually given in Hz [31].

Light is essential for the photopolymerisation process, the range of wavelengths used in photopolymerisation reactions usually entails both ultraviolet (UV) light (200 – 400 nm, UVA: 320 – 400 nm, UVB: 290 – 320 nm, and UVC: 190 – 290 nm) and visible light (400 – 700 nm). Infrared (IR) light (700 – 1000 nm) may also be used but it is not very common [32].

Since light is an electromagnetic wave, one of its significant photochemical properties is that it can transport energy. In order to exert an effect, the energy that a photon can transmit must be absorbed as well. When absorption of light by a

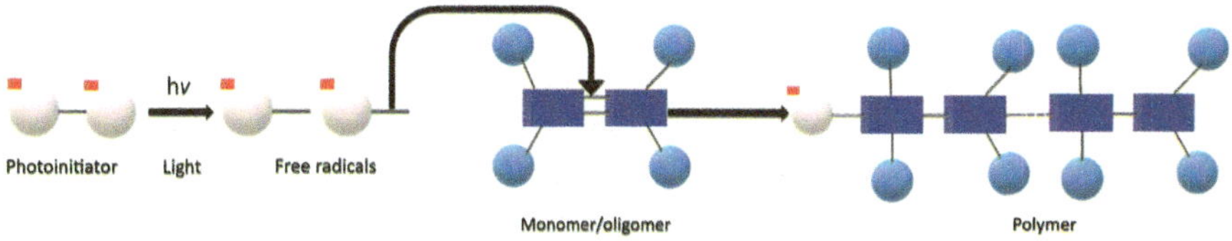

Fig. 4.8 General mechanism of photopolymerisation

molecule occurs it gets excited to a higher electronic energy level which then triggers a reaction (like dissociation or interaction) with other molecules [31].

A light beam can transport a certain amount of energy (in J cm^{-2} s^{-1}) that corresponds to the power density (in W cm^{-2}) and its given by $E_{beam}(\lambda_0) = I_0\ (\lambda_0)\ h\nu_0$. For laser sources $I_0\ (\lambda_0)$ is commonly referred as the number of photons per time unit in an area corresponding to the laser spot size, hence $E_{beam}(\lambda_0)$ expressed in J s^{-1} is equivalent to power P (given in W).

Energy sources used in photopolymerisation reactions include polychromatic (a filter is usually needed to narrow the wavelength range of the light being emitted) and monochromatic lights. Some examples are xenon lamps, mercury arc lamps, microwave lamps, light emitting diodes (LEDs) and lasers [32].

Lasers are a source of monochromatic light, the word laser itself is an acronym for light amplification by stimulated emission of radiation [33]. Laser beams offer multiple advantages including a wide selection of specific wavelengths, a high power concentrated in a small and accurately focalised spot which enables a short exposure time with high resolution.

2.2.2 Photoinitiators

As mentioned, in addition to a light source and a photopolymerisable monomer/oligomer, a photoinitiator is required following exposure to light as the light absorption of the monomer/oligomer itself does not yield enough energy to start the polymerisation process [31]. A photoinitiator is a compound (usually organic) that is thermally stable and capable of absorbing light with relatively high absorption coefficients in the UV and/or visible wavelength ranges [34].

Depending on the nature of the photoinitiator used, polymerisation reactions can be either radical or ionic (cationic and anionic), the type of initiating species also depends on the type of photoinitiator used [34]. However, cationic and anionic photopolymerization are noticeably less used hence we will focus on free radical mediated reactions.

Two types of compounds are employed as photoinitiators for free radical polymerizations which differ in their mode of action for generating reactive free radicals. Type I initiators undergo a very rapid bond cleavage after absorption of a photon. On the other hand, type II initiators form relatively long-lived excited triplet states capable of undergoing hydrogen-abstraction or electron-transfer

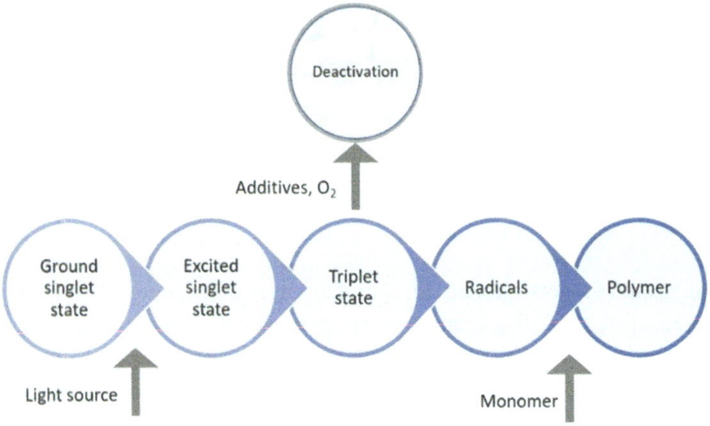

Fig. 4.9 Schematic representation of a free-radical photopolymerisation process

reactions with co-initiator molecules that are deliberately added to the monomer containing system [34].

The range at which a PI is capable to absorb is of high importance since it has to match the emission range of the light source. When a type II PI is exposed to light of an appropriate wavelength, it becomes excited and promoted from its ground singlet state (S_0) to its first excited singlet state (S_1) and then converted into its transient triplet state (T_1) which usually is the state yielding reactive radicals (R●). These radicals are then able to attack a monomer and start the polymerisation reaction [35] (Fig. 4.9).

Certain factors can also lead to deactivation of the T_1 state. These factors include monomer additives quenching reactions by oxygen and relaxation processes of the spectroscopic states. The efficiency of polymerization is determined by the experimental conditions including the characteristics of the light source (emission wavelength, number of photons emitted), the monomer concentration, the PI concentration and the presence of additives and oxygen [35].

A photoinitiator may be used alone or in conjunction with another molecule such as a photosensitizer (PS) or a co-initiator (CI). A PS is a molecule that absorbs light and transfers the excess of energy produced to the photoinitiator whereas a CI does not absorb light, but reacts with the photoinitiator to produce the initiating species.

A PS can act in two ways; 1) by transferring energy and 2) by transferring electrons. The most common type of PS are various aromatic compounds including benzophenone, acetophenone, triphenylamine, pyrene, naphthalene and anthracene [36].

There are many compounds that can act as free-radical photoinitiators. These types of photoinitiators offer many advantages such as low energy demand, room temperature reaction and fast processing. Some examples include aromatic ketones and acyl and bis(acyl)phosphine oxides [37].

Ideally, all PI systems must comply with different characteristics to make them appropriate for their use. These characteristics include good light absorption, good solubility/compatibility with the pre-polymer, good shelf-stability, high chemical reactivity with the initiating species, low sensitivity to oxygen inhibition, no influence on the final properties, excellent safety and biocompatibility.

A good photoinitiating system should possess good solubility or compatibility with the monomer/oligomer, it should be safe and it should have no effect on the final properties of the polymerized material. Added to those characteristics, a photoinitiator should essentially have a high reactivity (leading to fast polymerisation), in order to achieve fast polymerisation, an optimal absorption in the required spectral range is needed plus and an excellent photochemical reactivity, as well as a high chemical reactivity of the radicals.

However, for biomedical purposes, there are certain restrictions that hinder the use of photoinitiators with UV or visible absorption. These photoinitiators usually have limited water solubility and cytotoxicity problems [38] which limit their suitable range.

Some photoinitiators that have been evaluated for their use in the fabrication of scaffolds or drug delivery devices for biomedical research include;

Irgacure 2959: 2-hydroxy-1-[4-(hydroxyethoxy)phenyl]-2-methyl-1-propanone [39–42]
Irgacure 184: 1-hydroxycyclohexyl-1-phenyl ketone [43–45]
Irgacure 651: 2,2-dimethoxy-2-phenylacetophenone [46]
Irgacure 819: Bis(2,4,6-trimethylbenzoyl)-phenylphosphineoxide [47]
Riboflavin/triethanolamine [48–50]
Riboflavin/L-arginine [51]

The above mentioned photoinitiators have proved to be more biocompatible than others. Among the commercially available photoinitiators, Irgacure 2959 is the one with best biocompatible properties. In a study comparing the effect of 3 different photoinitiators (Irgacure 184, Irgacure 651 and Irgacure 2959) on 6 different cell lines, it was shown that Irgacure 2959 was best tolerated by 6 different cell lines. Toxicity increased as the concentrations of the initiator increased and the time of exposure to UV light, consequently increased as well [52].

Dye molecules can also be used as photoinititators, which in an electronically excited state, are capable of undergoing electron-transfer reactions with appropriate compounds (CIs). The free radical ions formed by electron transfer or by the decomposition of the radical ions, can initiate the polymerisation of monomers. In principle, the excited dye molecule can be reduced or oxidized, i.e. it can either accept or transfer an electron from or to the co-initiator [34].

One example of this is the photoinitiator system formed by riboflavin (vitamin B2, a naturally occurring pigment) and triethanolamine. The polymerisation mediated by this system occurs from the active radicals produced by the riboflavin triplet form interacting with the amine [53].

This system has a high absorption of energy at emissions at 312, 365 and 405 nm and more importantly, this system has shown to be less genotoxic and have a higher biocompatibility than other commercial PIs [51] including Irgacure 2959 [46]. This vitamin-amine system has been successfully used for photopolymerising polymers like 2-hydroxyethyl methacrylate (HEMA) [48, 51, 53–55] and polyethylene glycol diacrylate (PEGda) [46].

As an alternative for the use of triethanolamine, a completely biocompatible molecule such as an amino acid may be used. Arginine has been used as a co-initiator along with riboflavin (acting as a PI) for the fabrication of dextran-methacrylate hydrogels under a fluorescent lamp [51].

3 Materials and Biomedical uses of Sterolithography

In the biomedical area, it is essential for polymers to be biocompatible and safe for their clinical use. Thus, polymers have to be non-immunogenic, usually with good water solubility and non-toxic [56]. For their use in SLA in specific, pre-polymers must have photocurable moieties to make them suitable for fast polymerisation. These moieties are usually acrylates or fumarates [57].

However, this requirement for reactive groups in the molecules represents the biggest limitation for the use of stereolithography in biomedical purposes as there is a limited number of biocompatible and biodegradable photopolymerisable polymers. The reason for this is that although many crosslinked polymers are not cytotoxic, the unreacted monomer and photo-initiator residues entrapped within the crosslinked networks can be [58]. Therefore, it is of high importance to remove them to acceptable levels after the photopolymerisation reaction.

In the biomedical area, SLA has been widely studied for its use in tissue engineering for fabricating scaffolds that mimic physiological structures to deliver different types of molecules including cells, drugs and genes into the body. In spite of being extensively used for tissue engineering purposes, mainly due to its high accuracy for fabricating complex structures, research in the pharmaceutical area is still very limited, perhaps due to the above-mentioned material limitations.

In tissue engineering, there are many examples of the use of SLA with very promising results. Some examples include the use of poly(propylene fumarate) (PPF) to fabricate scaffolds for cell culture using microstereolithography. The scaffolds showed similar mechanical properties to those of human trabecular bone. Additionally, cultured fibroblasts were able to adhere to the scaffolds. PPF is biodegradable and its degradation products (fumaric acid and propylene glycol) are further biocompatible [26].

Another example of a polymer used is poly(ethylene glycol) (PEG) which has been widely implemented in biomedical applications due to its low toxicity and low interaction with proteins and cells [59]. PEG is an FDA approved material for diverse medical applications [60] which can be converted into a photo-crosslinkable molecule by attaching acrylates or methacrylates to its ends [61].

Poly(ethylene glycol) diacrylate (PEGda) has commonly been used for diverse biomedical applications [62] such as the fabrication of scaffolds for tissue engineering [63], capable of supporting cell encapsulation and proliferation. Furthermore, this polymer has shown good mechanical and biocompatible properties [46, 49].

SLA allows ease and accurate fabrication of hydrogels which have been widely studied for controlled release purposes. Vehse et al. fabricated UV-photopolymerised acetylsalicylic acid-loaded hydrogels (1 wt-%, 2 wt-% and 3 wt-%) to be used as scaffolds with Diode Laser Curing (DLC)-micro-stereolithography [64]. Likewise, bovine serum albumin photopolymerised hydrogel spheres were fabricated using PEGda, showing the tunability of the hydrogel properties to modify the protein release. The spheres showed to be potentially useful for the sustained delivery of proteins [65].

Arcaute et al. fabricated complex structures with multimaterials using a modified commercial SLA printer. They achieved a resolution of 500 μm, the purpose was to incorporate different materials for "spatially controlled" characteristics which can be useful for tissue engineering purposes. Moreover, the viability of human dermal fibroblasts encapsulated in the hydrogels was kept after fabrication [66].

Besides achieving a high resolution, as mentioned in previous examples, there is also a great possibility of encapsulating cells within the photopolymerised matrices. The first study encapsulating cells (Chinese hamster ovary cells) using stereolithography was done by Dhariwala et al. using poly (ethylene oxide) (PEO) along with PEG dimethacrylate to produce hydrogel scaffolds. They considered the SLA technology to be quite useful to fabricate complex tissue-like hydrogel structures and the cells showed to be viable and proliferating after 48 h post-fabrication [67].

Stereolithography has shown potential to fabricate structures that simulate tissues. Maling Gou fabricated 3D printed poly(ethylene glycol) diacrylate (PEGda) matrices by means of dynamic optical projection stereolithography to simulate liver tissue structure for detoxification using polydiacetylene nanoparticles installed to the hydrogel [68].

Mapili et al. fabricated poly (ethylene glycol) dimethacrylate (PEGdma) scaffolds with precise internal geometries by means of stereolithography. They incorporated microparticles to demonstrate the capability to include controlled-release carriers within the scaffold and the ability of stereolithography to create a single structure featuring multiple types of growth factors [69].

Recently, Placone et al. fabricated keratin hydrogels using the DLP technology. These scaffolds are reported to be fabricated using keratin in a UV-lithography based system for the first time. These hydrogels showed minimum cytotoxic interaction with cultured L929 cells and highlight the potential to use natural-derived polymers in photocrosslinkable resins for additive manufacturing [70].

Regarding medical devices, SLA has also proved to be suitable for producing personalised anti-acne devices (Fig. 4.10) containing salicylic acid. SLA showed better results for drug loading and drug diffusion when compared with FDM 3DP [7].

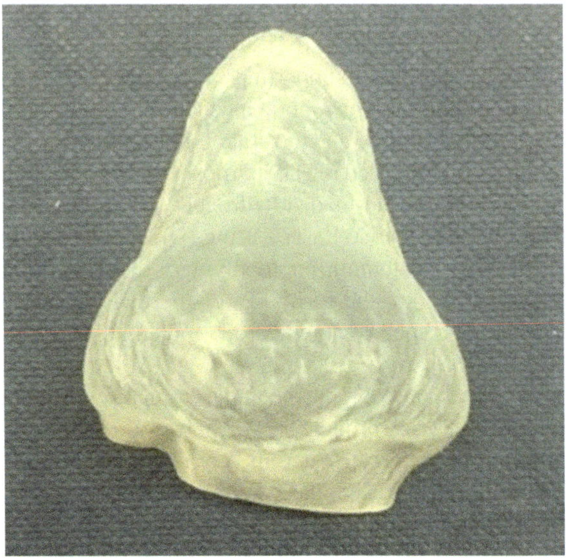

Fig. 4.10 3D printed personalised nose-shaped containing salicylic acid device fabricated with an SLA 3D printer. (Reprinted with permission from [71])

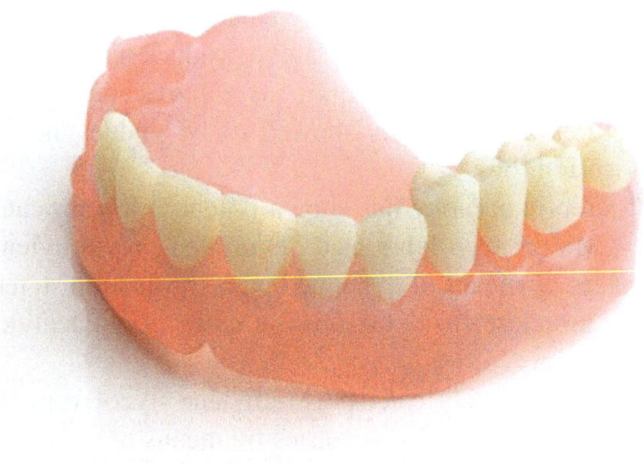

Fig. 4.11 3D printed dental prothesis using Dentca©. (Reprinted with permission from [73])

Furthermore, SLA has been extensively explored in the dental field for the fabrication of prostheses and restorative procedures. Importantly, there are FDA-approved 3D printed dental devices currently in the market. The first FDA approved device (Dentca™ Denture Base II) is a "light-curable resin indicated for fabrication and repair of full and partial removable dentures and baseplates" which was approved in 2015 [72].

Dentca™ denture base was approved as a system for the fabrication of dental prosthetics (Fig. 4.11) comprising a computer-aided design and manufacturing (CAD/CAM), digital denture base files based on a digital impression, a stereolithographic additive printer and a curing light equipment [74].

Similarly, NextDent™ Denture was approved by the FDA as a class-II material for 3D printing denture bases. The device is described as a light-cure resin, provided

Fig. 4.12 Hearing aid devices produced with the E-Shell 300 3D printer material. (Reprinted with permission from [77])

in a container indicated for the fabrication of denture bases fabricated in dental laboratories [75].

Another extensive use of SLA in the fabrication of medical devices focuses in hearing aids, in fact, the majority of the manufactured hearing aids are 3D printed using stereolithography technologies [76]. SLA is preferred due to its high resolution achieving a high level of detail and a smooth surface which is beneficial for fit and patient comfort. EnvisionTec is a leading company with more than 16 biomedically approved materials for producing hearing aid devices (Fig. 4.12) [77].

It is worth mentioning that the FDA usually approves finished medical devices and not individual materials for another use than the specified. The materials are assessed on their safety and effectiveness for that particular device. If a manufacturer wants to use an approved material for a different purpose, it would have to be evaluated by the FDA as a different classification regulation and the properties of the material will have to be evaluated for the new intended use (Food and Drug Administration, 2016).

The FDA evaluates and regulates this new technology by identifying similarities and differences with existing ones. It also has released a draft guidance for the "Technical Considerations for Additive Manufactured Devices – Draft Guidance for Industry and Food and Drug Administration Staff". This guidance includes the aspects to consider when producing a medical device by additive manufacturing, further discussed in Chaps. 9 and 11 of this book [78].

In pharmacoprinting, the approval of the first 3D printed drug for immediate release (Spritam®) in August 2015 opened a thrilling new chapter in pharmaceutical manufacturing for oral dosage forms. One of the main objectives of pharmacoprinting is to produce personalised medicines on demand and most of the research is focused on controlled and targeted release [79–82]. For the approval of this products, they should also be manufactured complying existing standards set in relevant guidance, including the 21 CFR 200 s and 300 s [78].

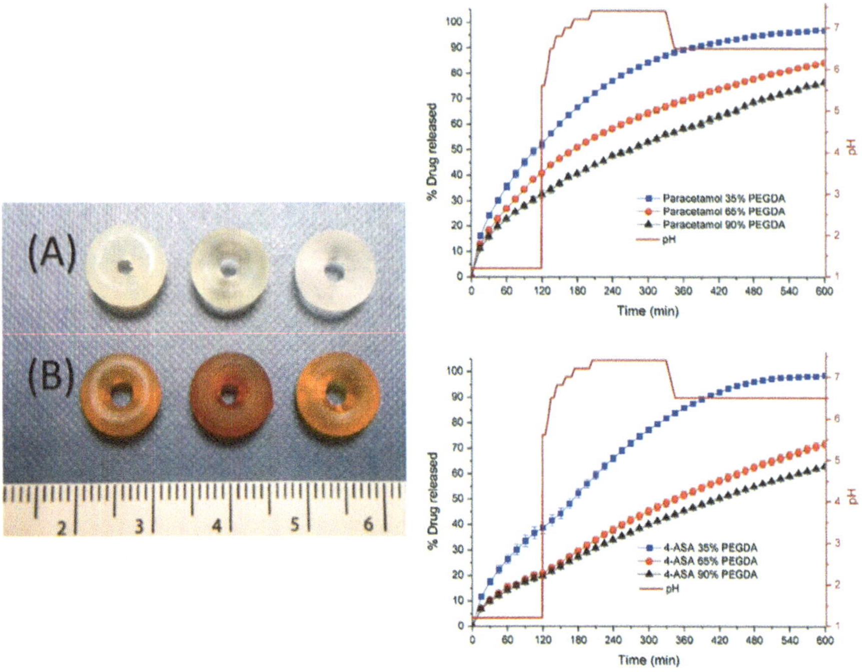

Fig. 4.13 SLA 3D printed Printlets containing paracetamol (top left) and 4-ASA (bottom left). Drug release profiles from them (right). (Reprinted with permission from [71])

Research in the field of oral drug delivery using SLA is still very limited. One of the limited papers available in the literature is by Wang et al. who fabricated 4-aminosalicylic acid (4-ASA) and paracetamol loaded tablets showing no drug degradation which is a definite advantage when compared with a previous study performed using FDM [8]. Figure 4.13 shows an image of 3D printed tablets (herein known as printlets: a term that we have coined to describe 3DP oral dosage forms) with a noticeable good resolution even when producing a complex shape (torus), which would be very difficult to achieve using traditional manufacturing methods [80]. The drug release profiles from the torus-shaped tablets showed cross-linking degree dependence [71].

In another study, SLA was used to fabricate ibuprofen-loaded hydrogels (Fig. 4.14). This proof of concept project showed the feasibility to use in a commercial desktop 3D printer, a non-commercial water-soluble photoinitiating system consisting of riboflavin and triethanolamine. In this project, the use of different percentages of water in the initial formulation was also explored, showing that the crosslinking density is slightly modified as the water content increases (up to 30%). However, this dilution with water did not seem to significantly affect the speed at which the drug was released [50].

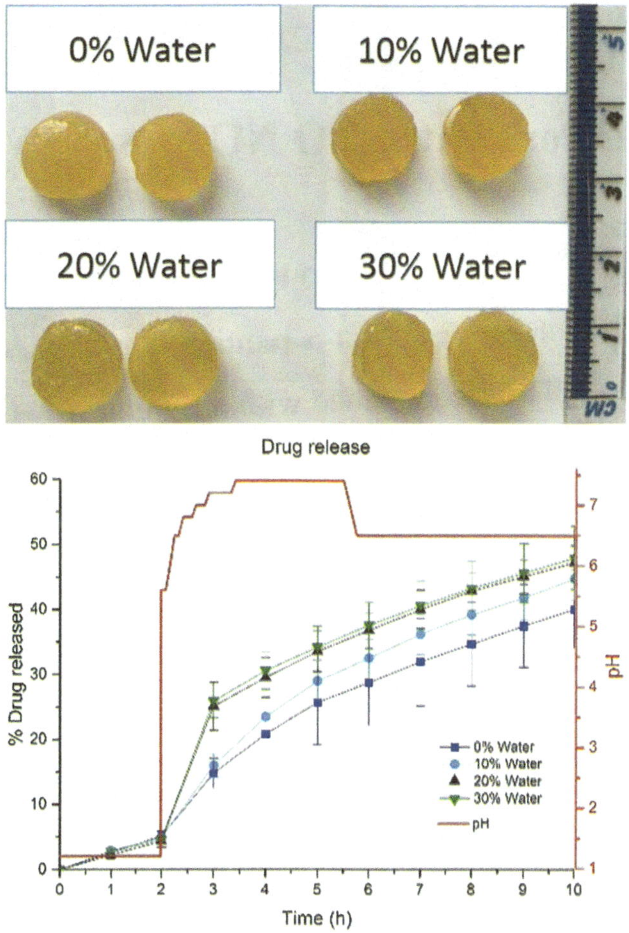

Fig. 4.14 SLA 3D printed hydrogels with different initial water content (left) and the drug release profiles for the 3DP hydrogels (right). The graph shows the speed of drug release is not greatly affected by the initial water content in the formulations. (Reprinted with permission from [50])

Recently, Larush et al. fabricated smart pH-responsive hydrogels using DLP to explore the effect that swelling and shape have on the release profile of a pH-independent dye, Sulforhodamine B. They used acrylic acid and PEGda as monomers which showed that the release was increased at higher pH, making them suitable for drug delivery in the intestine Cytotoxocity test in CaCo2 cells suggested no cytotoxicity effect from the 3D printed tablets [83]. Complex and precise microarchitechtural structures (Fig. 4.15) have been also fabricated using μSLA using PEG 200 and PEG 1000 as base polymers. Differences in erosion, drug loading and drug release were observed depending on the molecular weight of the base polymer [84].

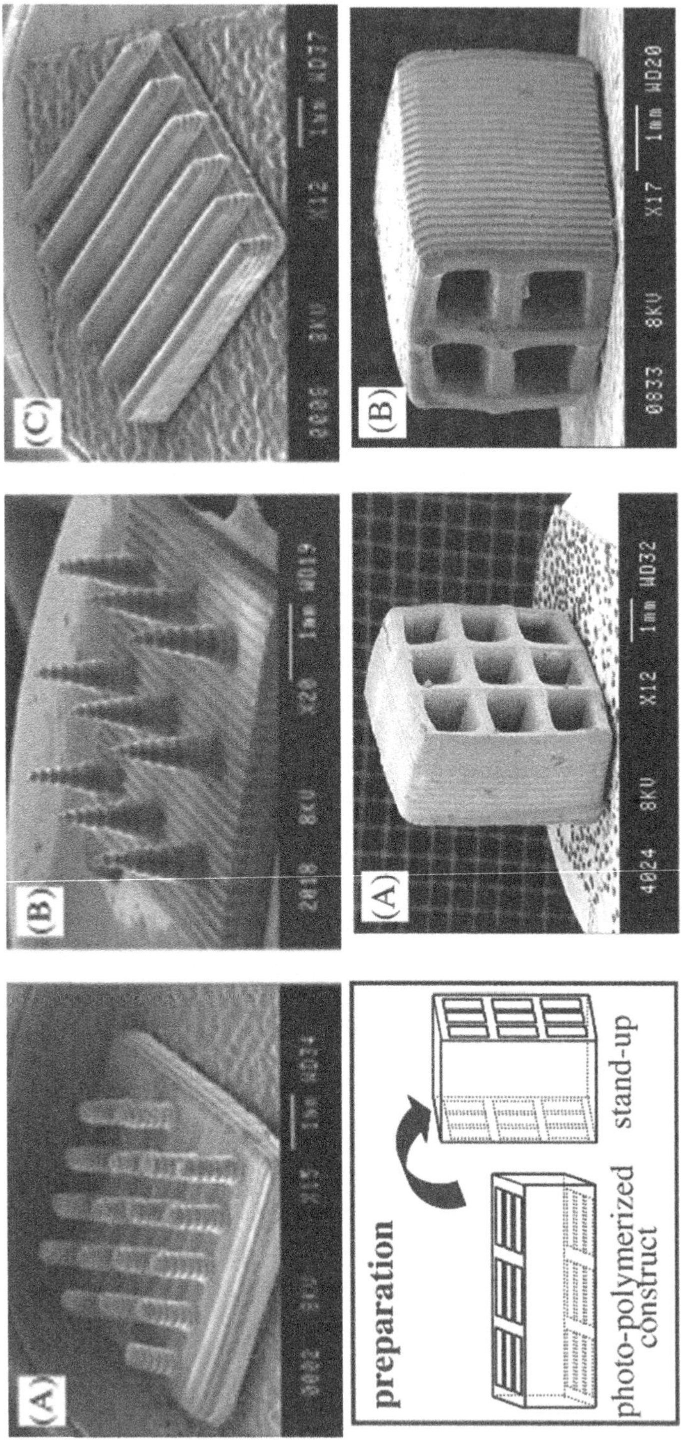

Fig. 4.15 SEM images of microstructures fabricated using μSLA. (Reprinted with permission from [84])

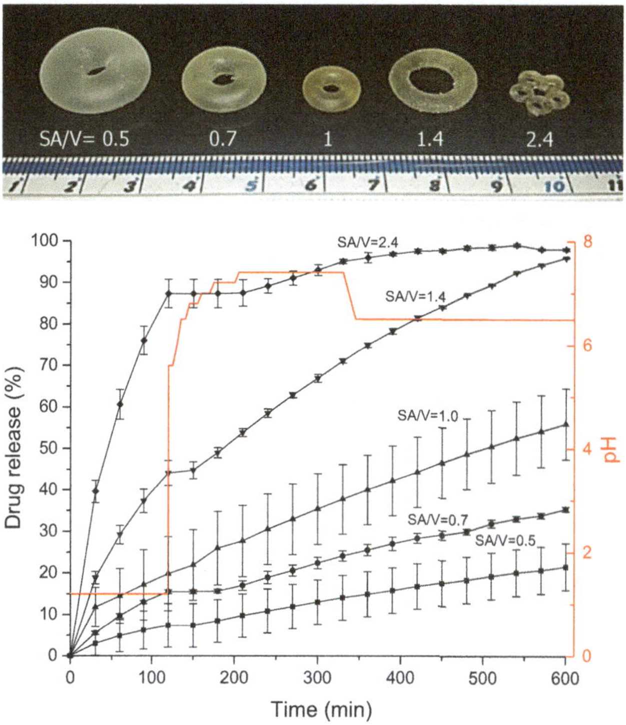

Fig. 4.16 3D printed tablets with different SA/V values (left) and their corresponding drug release profiles (right) [85]

It is clear that the high resolution of this 3DP technique allows the fabrication of drug delivery devices that were not possible using traditional manufacturing methods [80]. Similarly, a group of researchers in the UCL School of Pharmacy produced printlets with complex shapes (Fig. 4.16), demonstrating the influence that tablet dimensions have on drug release. The results show that the speed of drug release is greatly influenced by the surface-area-to-volume (SA/V) ratio of the printlet. This can be useful to modify the dose and the shape of tablets without altering the drug release profile, as well as to duplicate drug release profiles without changing the formulation.

Similarly, transdermal drug delivery could be potentially benefited by the use of stereolithography. SLA has been explored for the production of microneedles, considered to be less traumatic and less painful than traditional needles. They can also be a convenient replacement for hypodermic injections, making the administration of a diverse number of molecules (including biomacromolecules) patient-friendly, increasing patience compliance. Compared to other fabrication methods of microneedles, 3DP represents the easiest method for producing large batches with complex structures [86].

Matsuda and Mizutani fabricated indomethacin loaded microneedles that proved to slow-release the drug, allowing a favourable effect when tested in rat tissue. The microneedles were fabricated using biodegradable photocurable copolymers and a customised SLA apparatus [87].

3.1 Biocompatibility

As aforementioned, uncured monomers or unreacted PI molecules are not usually biocompatible, hence, researchers are striving to identify new biocompatible materials and to evaluate and reduce the risk of cytotoxic effects of existing photocurable resins. A logical approach to avoid the effect that residual materials can have is by removing them after printing. Popov et al., found that monomers, low molecular weight oligomers and other toxic remainders can be removed from olygocarbonatedimethacrylate-hydroxyapatite composite materials using a supercritical carbon dioxide treatment [88]. Similarly, Inoue and Ikuta reported that cytotoxicity from a polymer produced using microstereolithography could be removed by sublimation of the unreacted materials when the polymer was treated. This could be achieved using high temperatures in a nitrogen atmosphere [89].

Likewise, the safety of photocured materials can be assessed when they are used for producing cell-containing matrices or scaffolds. Mazzoccoli et al. found that cell viability in crosslinked hydrogels depends on the molecular weight and concentration of the pre-polymer used. In their study, viability of cells from human epithelial origin was nearly 90% when the molecular weight of PEGda was higher (3400, compared to 400) and when the total polymer concentration was around 20%. The detrimental effect of uncured monomers and the PI can be minimised by using a low concentration of the photopolymerisable pre-polymer and high molecular weight pre-polymers [90].

Post-fabrication UV exposure may also be useful to remove the presence of unreacted moieties. Since resin-solidification occurs during 3D printing at UV doses below the level required for full conversion, the post-fabrication process may not be complete. However, conversion can be completed by an intense exposure to UV light ('post-cure' process). Moreover, post-curing can also be useful to improve the stability of 3D printed objects by enhancing their physicochemical properties [6].

4 Conclusion

Stereolithography is a 3D printing technique that possess a great potential for manufacturing highly complex internal and external geometries with an outstanding resolution. It has been widely used in different manufacturing areas such as aerospace, automotive and fashion design. In the biomedical field, it has provided a very useful method to produce tissue-like structures and scaffolds. However, in the pharmaceutical area, research is still very limited perhaps due to safety limitations of the currently available photopolymerisable monomers. However, the exploration of potential new applications and new safer materials is constantly growing, and perhaps complete biocompatibility of the materials used may be achieved in the near future.

References

1. Hull, C. W. Apparatus for production of three-dimensional objects by stereolithography. US Patent 4,575,330 1–16 (1986). https://doi.org/10.1145/634067.634234.
2. Chua CK, Leong KF, An J. Introduction to rapid prototyping of biomaterials. Rapid Prototyp Biomater Princ Appl. 2014:1–15. https://doi.org/10.1533/9780857097217.1.
3. Kruth JP. Material Incress manufacturing by rapid prototyping techniques. CIRP Ann Manuf Technol. 1991;40:603–14.
4. Pham D, Gault R. A comparison of rapid prototyping technologies. Int J Mach Tools Manuf. 1998;38:1257–87.
5. Gardan J. Additive manufacturing technologies: state of the art and trends. Int J Prod Res. 2016;54:3118–32. https://doi.org/10.1080/00207543.2015.1115909.
6. Vitale A, Cabral JT. Frontal conversion and uniformity in 3D printing by photopolymerisation. Materials (Basel). 2016;9:760–72.
7. Goyanes A, Det-Amornrat U, Wang J, Basit AW, Gaisford S. 3D scanning and 3D printing as innovative technologies for fabricating personalized topical drug delivery systems. J Control Release. 2016;234:41–8.
8. Goyanes A, Buanz ABM, Hatton GB, Gaisford S, Basit AW. 3D printing of modified-release aminosalicylate (4-ASA and 5-ASA) tablets. Eur J Pharm Biopharm. 2015;89:157–62.
9. Stansbury JW, Idacavage MJ. 3D printing with polymers: challenges among expanding options and opportunities. Dent Mater. 2016;32:54–64.
10. Koslow T. 11 Best Resin (DLP/SLA) 3D printers in 2017 | All3DP. April 12 2017. Available at: https://all3dp.com/1/best-resin-dlp-sla-3d-printer-kit-stereolithography/#uncia-3d. Accessed 19 Oct 2017.
11. Systems, 3D. 3D systems: our story. 2013: 1–7. Available at: https://www.3dsystems.com/our-story?smtNoRedir=1&_ga=2.188103467.2074466491.1503585626-2013118968.1503585626. Accessed 24 Aug 2017.
12. Kinematics Fold | Nervous System | Somerville. Available at: https://n-e-r-v-o-u-s.com/projects/albums/kinematics-fold/content/video-kinematics-fold/. Accessed 28 Mar 2018.
13. European Patent Office. Charles W. Hull (USA). Available at: https://www.epo.org/learning-events/european-inventor/finalists/2014/hull.html. Accessed 4 Sep 2017.
14. Markillie P. A third industrial revolution | the economist. De Economist. 2012;1 Available at: http://www.economist.com/node/21552901. Accessed 14 Oct 2017.
15. Hegde M, et al. 3D printing all-aromatic polyimides using mask-projection stereolithography: processing the nonprocessable. Adv Mater. 2017;29:1701240.
16. Shapeways. WEB BANGLE (6MQRWHKA7) by JAXJEWELRY. Available at: https://www.shapeways.com/product/6MQRWHKA7/web-bangle?optionId=43536683&li=marketplace. Accessed 6 Feb 2018.
17. 3D Architech | Under Armour | US. Available at: https://www.underarmour.com/en-us/3d-architech. Accessed 12 Sept 2017.
18. 3ders.org – Chinese city of Nanjing gets first 3D printed bridge railings | 3D Printer News & 3D Printing News. Available at: http://www.3ders.org/articles/20170821-chinese-city-of-nanjing-gets-first-3d-printed-bridge-railings.html. Accessed 30 Aug 2017.

19. Keating SJ, Leland JC, Cai L, Oxman N. Toward site-specific and self-sufficient robotic fabrication on architectural scales. Sci Robot. 2017;2:eaam8986.
20. 3ders.org – Russian spacewalking cosmonauts release world's first 3D printed satellite from ISS | 3D Printer News & 3D Printing News. Available at: http://www.3ders.org/articles/20170817-russian-spacewalking-cosmonauts-release-worlds-first-3d-printed-satellite-from-iss.html. Accessed 30 Aug 2017.
21. Jang J, et al. 3D printed complex tissue construct using stem cell-laden decellularized extracellular matrix bioinks for cardiac repair. Biomaterials. 2017;112:264–74.
22. Graham AD, et al. High-resolution patterned cellular constructs by droplet-based 3D printing. Sci Rep. 2017;7:7004.
23. Atkins T, Escudier M. A dictionary of mechanical engineering: Oxford University Press; 2013.
24. Szilvśi-Nagy M, Mátyási G. Analysis of STL files. Math Comput Model. 2003;38:945–60.
25. Iancu C, Iancu D, Stăncioiu A. From CAD model to 3D print via' STL' file format. Fiability Durab/Fiabilitate si Durab. 2010;1:73–80.
26. Jin WL, Phung XL, Kim B, Lim G, Cho DW. Fabrication and characteristic analysis of a poly(propylene fumarate) scaffold using micro-stereolithography technology. J Biomed Mater Res Part B Appl Biomater. 2008;87:1–9.
27. Tumbleston JR, et al. Continuous liquid interface production of 3D objects. Science. 2015;347:1349–52.
28. Tehfe M, Louradour F, Lalevée J, Fouassier J-P. Photopolymerization reactions: on the way to a green and sustainable chemistry. Appl Sci. 2013;3:490–514.
29. Fouassier JP, Lalevée J. Photoinitiators for polymer synthesis. Weinheim: Wiley-VCH Verlag GmbH & Co. KGaA; 2012. p. 3–9. https://doi.org/10.1002/9783527648245.ch1.
30. Raman R, Bashir R. Chapter 6 – Stereolithographic 3D bioprinting for biomedical applications. In: Essentials of 3D biofabrication and translation. San Diego: Elsevier; 2015. https://doi.org/10.1016/B978-0-12-800972-7.00006-2.
31. Bailey RA, et al. Chemistry of the environment. San Diego: Elsevier; 2002. p. 73–90. https://doi.org/10.1016/B978-012073461-0/50051-X.
32. Fouassier JP, Lalevée J. Photoinitiators for polymer synthesis. Weinheim: Wiley-VCH Verlag GmbH & Co. KGaA; 2012. p. 11–20. https://doi.org/10.1002/9783527648245.ch2.
33. Milonni PW, Eberly JH. Laser physics. New York: Wiley; 2010. p. 1–15. https://doi.org/10.1002/9780470409718.ch1.
34. Schnabel W. Polymers and light – fundamentals and technical applications. Weinheim: Wiley-VCH Verlag GmbH & Co. KGaA; 2007. https://doi.org/10.1002/9783527611027.
35. Fouassier JP, Allonas X, Laleve J, Dietlin C. Photochemistry and photophysics of polymer materials. Hoboken: Wiley; 2010. p. 351–419. https://doi.org/10.1002/9780470594179.ch10.
36. Ravve A. Light-associated reactions of synthetic polymers. New York: Springer; 2006. p. 23–122. https://doi.org/10.1007/0-387-36414-5_2.
37. Colley CS, et al. Probing the reactivity of photoinitiators for free radical polymerization: time-resolved infrared spectroscopic study of benzoyl radicals. J Am Chem Soc. 2002;124:14952–8.
38. Tseng S-J, et al. Controlled hydrogel photopolymerization inside live systems by X-ray irradiation. Soft Matter. 2012;8:1420–7. https://doi.org/10.1039/c1sm06682j.
39. Zhong C, Wu J, Reinhart-King CA, Chu CC. Synthesis, characterization and cytotoxicity of photo-crosslinked maleic chitosan-polyethylene glycol diacrylate hybrid hydrogels. Acta Biomater. 2010;6:3908–18.
40. Melchels FPW, Feijen J, Grijpma DW. A poly(d,l-lactide) resin for the preparation of tissue engineering scaffolds by stereolithography. Biomaterials. 2009;30:3801–9. https://doi.org/10.1016/j.biomaterials.2009.03.055.
41. Nuttelman CR, Henry SM, Anseth KS. Synthesis and characterization of photocrosslinkable, degradable poly(vinyl alcohol)-based tissue engineering scaffolds. Biomaterials. 2002;23:3617–26.
42. Leach JB, Bivens KA, Patrick CW, Schmidt CE. Photocrosslinked hyaluronic acid hydrogels: natural, biodegradable tissue engineering scaffolds. Biotechnol Bioeng. 2003;82:578–89. https://doi.org/10.1002/bit.10605.

43. Torres-Lugo M, Peppas NA. Molecular design and in vitro studies of novel pH-sensitive hydrogels for the oral delivery of calcitonin. Macromolecules. 1999;32:6646–51.
44. Cevik O, Gidon D, Kizilel S. Visible-light-induced synthesis of pH-responsive composite hydrogels for controlled delivery of the anticonvulsant drug pregabalin. Acta Biomater. 2015;11:151–61.
45. Fairbanks BD, Schwartz MP, Bowman CN, Anseth KS. Photoinitiated polymerization of PEG-diacrylate with lithium phenyl-2, 4, 6- trimethylbenzoylphosphinate: polymerization rate and cytocompatibility. Biomaterials. 2009;30:6702–7.
46. Nguyen AK, et al. Two-photon polymerization of polyethylene glycol diacrylate scaffolds with riboflavin and triethanolamine used as a water-soluble photoinitiator. Regen Med. 2013;8:725–38.
47. Kang H-W, Cho D-W. Development of an indirect stereolithography technology for scaffold fabrication with a wide range of biomaterial selectivity. Tissue Eng Part C Methods. 2012;18:719–29.
48. Ahmad I, et al. Photoinitiated polymerization of 2-hydroxyethyl methacrylate by riboflavin/triethanolamine in aqueous solution: a kinetic study. ISRN Pharm. 2013;2013:958712.
49. Nguyen KT, West JL. Photopolymerizable hydrogels for tissue engineering applications. Biomaterials. 2002;23:4307–14.
50. Martinez PR, Goyanes A, Basit AW, Gaisford S. Fabrication of drug-loaded hydrogels with stereolithographic 3D printing. Int J Pharm. 2017;532:313–7. https://doi.org/10.1016/j.ijpharm.2017.09.003.
51. Kim S, Chu C-C. Visible light induced dextran-methacrylate hydrogel formation using (−)-riboflavin vitamin B2 as a photoinitiator and L-arginine as a co-initiator. Fibers Polym. 2009;10:14–20.
52. Williams CG, Malik AN, Kim TK, Manson PN, Elisseeff JH. Variable cytocompatibility of six cell lines with photoinitiators used for polymerizing hydrogels and cell encapsulation. Biomaterials. 2005;26:1211–8.
53. Orellana B, Rufs AM, Encinas MV, Previtali CM, Bertolotti S. The photoinitiation mechanism of vinyl polymerization by riboflavin/triethanolamine in aqueous medium. Macromolecules. 1999;32:6570–3.
54. Bertolotti SG, Previtali CM, Rufs AM, Encinas MV. Riboflavin/triethanolamine as photoinitiator system of vinyl polymerization. A mechanistic study by laser flash photolysis. Macromolecules. 1999;32:2920–4.
55. Encinas MV, Rufs AM, Bertolotti S, Previtali CM. Free radical polymerization photoinitiated by riboflavin/amines. Effect of the amine structure. Macromolecules. 2001:2845–7. https://doi.org/10.1021/ma001649r.
56. Grund S, Bauer M, Fischer D. Polymers in drug delivery-state of the art and future trends. Adv Eng Mater. 2011;13:B61–87.
57. Chia HN, Wu BM. Recent advances in 3D printing of biomaterials. J Biol Eng. 2015;9:4.
58. Oskui SM, et al. Assessing and reducing the toxicity of 3D-printed parts. Environ Sci Technol Lett. 2016;3:1–6.
59. Wang N, et al. Synthesis of degradable functional poly(ethylene glycol) analogs as versatile drug delivery carriers. Macromol Biosci. 2007;7:1187–98.
60. Yu J, et al. In situ covalently cross-linked PEG hydrogel for ocular drug delivery applications. Int J Pharm. 2014;470:151–7.
61. Drury JL, Mooney DJ. Hydrogels for tissue engineering: scaffold design variables and applications. Biomaterials. 2003;24:4337–51.
62. Cruise GM, Scharp DS, Hubbell JA. Characterization of permeability and network structure of interfacially photopolymerized poly(ethylene glycol) diacrylate hydrogels. Biomaterials. 1998;19:1287–94.
63. Leach JB, Schmidt CE. Characterization of protein release from photocrosslinkable hyaluronic acid-polyethylene glycol hydrogel tissue engineering scaffolds. Biomaterials. 2005;26:125–35.

64. Vehse M, Petersen S, Sternberg K, Schmitz KP, Seitz H. Drug delivery from poly(ethylene glycol) diacrylate scaffolds produced by DLC based micro-stereolithography. Macromol Symp. 2014;346:43–7.
65. Mellott MB, Searcy K, Pishko MV. Release of protein from highly cross-linked hydrogels of poly(ethylene glycol) diacrylate fabricated by UV polymerization. Biomaterials. 2001;22:929–41.
66. Arcaute K, Mann BK, Wicker RB. Stereolithography of three-dimensional bioactive poly(ethylene glycol) constructs with encapsulated cells. Ann Biomed Eng. 2006;34:1429–41.
67. Dhariwala B, Hunt E, Boland T. Rapid prototyping of tissue-engineering constructs, using photopolymerizable hydrogels and stereolithography. Tissue Eng. 2004;10:1316–22.
68. Gou M, et al. Bio-inspired detoxification using 3D-printed hydrogel nanocomposites. Nat Commun. 2014;5:1–9.
69. Mapili G, Lu Y, Chen S, Roy K. Laser-layered microfabrication of spatially patterned functionalized tissue-engineering scaffolds. J Biomed Mater Res Part B Appl Biomater. 2005;75:414–24.
70. Placone JK, et al. Development and characterization of a 3D printed, keratin-based hydrogel. Ann Biomed Eng. 2017;45:237–48.
71. Wang J, Goyanes A, Gaisford S, Basit AW. Stereolithographic (SLA) 3D printing of oral modified-release dosage forms. Int J Pharm. 2016;503:207–12.
72. Food and Drug Administration. Department of health & human services. Approved Premarket Notification: Dentca Denture Base II. 510(k) number K160244. 2017. Available at: https://www.accessdata.fda.gov/cdrh_docs/pdf16/K162044.pdf. Accesed 24 Aug 2017.
73. DENTCA. DENTCA 3D Printed Denture | Dentca. (2016). Available at: https://www.dentca.com/products/dentca-3d. Accessed 22 Aug 2017.
74. Food and Drug Administration. Department of health & human services. Approved Premarket Notification: Dentca denture base. 510(k) Number K143033. 2015. Available at: https://www.accessdata.fda.gov/cdrh_docs/pdf14/K143033.pdf. Accessed 22 Aug 2017.
75. Food and Drug Administration. Department of health & human services. Approved Premarket Notification: NextDent™ Denture/E-Denture. 510(k) Number K162572. 2017. Available at: https://www.accessdata.fda.gov/cdrh_docs/pdf16/K162572.pdf. Accesed 24 Aug 2017.
76. Sharma R. The 3D printing revolution you have not heard about. In: Forbes.2013. Available at: https://www.forbes.com/sites/rakeshsharma/2013/07/08/the-3d-printing-revolution-you-have-not-heard-about/#33bacc131a6b. Accessed 1 Dec 2017.
77. ENVISIONTEC INC. 3D Printed Hearing Aid Shells, Molds, Inner-Ear Devices | EnvisionTEC. Available at: https://envisiontec.com/3d-printing-industries/medical/hearing-aid/. Accessed 1 Dec 2017.
78. Di Prima M, et al. Additively manufactured medical products – the FDA perspective. 3D Print Med. 2015;2:1.
79. Trenfield SJ, Awad A, Goyanes A, Gaisford S, Basit AW. 3D printing pharmaceuticals: drug development to frontline care. Trends Pharmacol Sci. 2018;39(5):440–51.
80. Goole J, Amighi K. 3D printing in pharmaceutics: a new tool for designing customized drug delivery systems. Int J Pharm. 2016;499(1–2):376–94.
81. Awad A, Trenfield SJ, Gaisford S, Basit AW. 3D printed medicines: A new branch of digital healthcare. Int J Pharm. 2018;548(1):586–96.
82. Awad A, Trenfield SJ, Goyanes A, Gaisford S, Basit AW. Reshaping drug development using 3D printing. Drug Discov Today. 2018; https://doi.org/10.1016/j.drudis.2018.05.025.
83. Larush L, Kaner I, Fluksman A, Tamsut A, Pawar AA. 3D printing of responsive hydrogels for drug-delivery systems. J 3D Print Med. 2017;1:219–29.
84. Kwon IK, Matsuda T. Photo-polymerized microarchitectural constructs prepared by microstereolithography (μSL) using liquid acrylate-end-capped trimethylene carbonate-based prepolymers. Biomaterials. 2005;26:1675–84.
85. Martinez PR, Goyanes A, Basit AW, Gaisford S. Influence of geometry on the drug release profiles of stereolithographic (SLA) 3D printed tablets. AAPS PharmSciTech. 2018; https://doi.org/10.1208/s12249-018-1075-3.

86. Miller PR, et al. Integrated carbon fiber electrodes within hollow polymer microneedles for transdermal electrochemical sensing. Biomicrofluidics. 2011;5:13415.
87. Matsuda T, Mizutani M. Liquid acrylate-endcapped biodegradable poly(epsilon-caprolactone-co-trimethylene carbonate). II. Computer-aided stereolithographic microarchitectural surface photoconstructs. J Biomed Mater Res. 2002;62:395–403.
88. Popov VK, et al. Laser stereolithography and supercritical fluid processing for custom-designed implant fabrication. J Mater Sci Mater Med. 2004;15:123–8.
89. Inoue Y, Ikuta K. Detoxification of the photoeurable polymer by heat treatment for microstereolithography. Procedia CIRP. 2013;5:115–8.
90. Mazzoccoli JP, Feke DL, Baskaran H, Pintauro PN. Mechanical and cell viability properties of crosslinked low- and high-molecular weight poly(ethylene glycol) diacrylate blends. J Biomed Mater Res A. 2010;93:558–66.

Chapter 5
Powder Bed Fusion: The Working Process, Current Applications and Opportunities

Fabrizio Fina, Simon Gaisford, and Abdul W. Basit

Abstract Powder bed fusion (PBF) is an umbrella term for three separate three-dimensional (3D) printing technologies; selective laser sintering (SLS), direct metal laser sintering (DMLS) and selective laser melting (SLM). These processes share the same printing procedure where powder particles are selectively fused by a local thermal process generated from a laser in a layer-by-layer manner. However, they differ in the materials used and energy transmitted. Thermoplastic polymers are often employed by SLS whereby a laser superficially connects the powder particles together by a process known as sintering. DMLS and SLM is still a sintering process although metals and powders are most often used. DMLS and SLM are mainly used in the aerospace, maritime and automotive industries although SLS has been successfully introduced into the medical arena for its applications in tissue engineering. In this chapter, we further explore how the SLS technology has proven its amenability in the printing and manufacture of pharmaceuticals.

Keywords Additive manufacturing · Personalized medicines · Bioprinting · Drug product manufacturing · Drug delivery systems · Pharmaceutics

F. Fina (✉)
Department of Pharmaceutics, UCL School of Pharmacy,
University College London, London, UK
e-mail: fabrizio.fina.14@ucl.ac.uk

S. Gaisford · A. W. Basit
Department of Pharmaceutics, UCL School of Pharmacy,
University College London, London, UK

FabRx Ltd., Ashford, Kent, UK

A. W. Basit, S. Gaisford (eds.), *3D Printing of Pharmaceuticals*,
AAPS Advances in the Pharmaceutical Sciences Series 31,
https://doi.org/10.1007/978-3-319-90755-0_5

1 Historical Development

Powder bed fusion (PBF) includes three main processes; (1) selective laser sintering, (2) direct metal laser sintering and (3) selective laser melting where a three-dimensional (3D) object can be printed in a layer-by-layer method from employing a powder bed and a laser (Fig. 5.1).

In 1984, a University of Texas student, Carl Deckard, invented the first PBF technique, now known as selective laser sintering (SLS) (Fig. 5.1) [1]. In conjunction with his supervisor, Professor Beaman, Deckard developed a machine that was capable of creating solid entities from the fusion of powder particles when accompanied with a powerful laser. At that time, Texas was experiencing a very prosperous moment from the expansion of its oil industry, and because of this, significant financial benefits were offered to universities to encourage the exploration of new technologies. As a result, Beaman was then able to provide Deckard with all the equipment needed to develop his invention. Deckard initially coined his method as a "part generation by layerwise selective sintering" (PGLSS) and was later modified to selective laser sintering (SLS). After 2 years of development, Deckard and Beaman decided to collaborate with a corporate partner, Nova Automation. In 1990, Deckard patented the SLS technology [2] and Nova Automation became the DMT ("desk top manufacturing") Corporation. In their patent, the apparatus was described to employ a 100 W Nd:YAG laser (less expensive than a CO_2 laser) with acrylonitrile butadiene styrene (ABS) powder [1]. Although the invention was not restricted to polymer powders, no description of metal powders was reported in the patent. Prior to the development of SLS, mechanical engineers were taught to keep their designs as simple as possible when creating new parts due to production machinery limitations. SLS, according to Beaman, would revolutionise the way of thinking for the next generation engineers. Complexity would then be something to embrace and no longer avoided.

Fig. 5.1 Three main PBF technologies

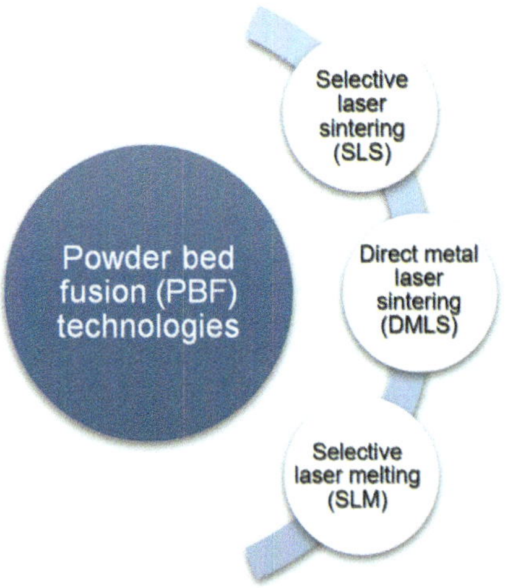

The second PBF technique was in development in 1990 when the University of Texas was exploring the application of SLS technology to metals. They named the process as Direct Metal Laser Sintering (DMLS) although the first attempts of using zinc or lead were not successful. However in 1994, Fraunhofer Institute of Production Technology reported the first success using 316 L stainless steel powder [3].

By the end of 1995, a German research organisation, Fraunhofen Institute of Laser Technology, stretched the boundaries of SLS even further by developing and patenting a technology that could completely melt either metals or ceramics to create dense components [4]. This new PBF technology was then named Selective Laser Melting (SLM).

2 The Working Process

The overall printing procedure is the same for all three SLS, DMLS and SLM technologies [5]. Successive layers are connected through a thermal process ranging from sintering (the superficial connection of the powder particles) through different degrees of partial melting until a complete melting state is achieved. The energy required to print the object is supplied from a combination of a heating system of the printer and a laser beam. A beam deflection system based on galvano mirrors is used to direct the laser beam on to the powder bed surface following the parameters outlined in an .stl file.

All three processes use a single type of powder or a combination of powders as the starting material. However, each PBF technique differs according to the material(s) employed and for the external energy transmitted to the powder bed surface.

- SLS usually refers to thermoplastic polymers and the energy applied should only be enough to sinter the particles together.
- DMLS is also an example of a sintering process, however, metallic powder is chosen as the starting [6].
- SLM is usually applied when using metal powders or ceramics. The transmitted energy is often higher when compared with the former techniques as the powder is expected to melt completely.

Figure 5.2 shows the general printing process of PBF. A reservoir platform (right) is set at its lower position where the powder is poured before starting the printing process.

The building platform is set at its highest position. Powder is distributed at the top using a roller to prepare a flat and homogeneous powder bed where the printing process will take place. The printing process starts with the activation of the laser that sinters the first layer onto the powder bed. Subsequently, the reservoir platform moves up whilst the building platform descends. The roller will then uniformly distribute a thin layer of powder from the reservoir platform to the building platform. The laser is then activated again to sinter the second layer and the process is repeated

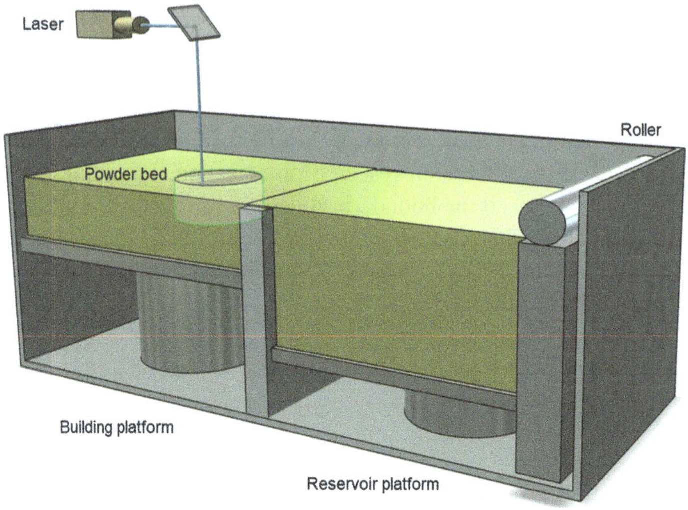

Fig. 5.2 Basic representation of the printing process for PBF. (Adapted with permission from [7])

layer-by-layer until completion. When the printing process is complete, the object is recovered underneath the powder bed. Excess powder on the object surface is then removed using compressed air or by brushing. When the object is clean, post-processing may be required to increase the strength of the object or to improve its final appearance. Coating with an appropriate material allows the enhancement of its tensile strength and the surface hardness while surface finishing can improve the surface precision.

Advantages of PBF printing include a solvent-free process, capable of creating complex 3D objects without printing additional reinforcement as the un-sintered powder itself acts as support. SLS is also able to produce objects with high controlled porosity and high pore connectivity [8]. One disadvantage is the large quantity of powder required to complete the printing process. The unsintered powder material, however, can be recycled and re-used and thus, limit the overall material wastage. Additionally, the surface of the final object is quite rough, therefore, a post-treatment process may be required. Processing temperatures for metals and ceramics when using SLM, however, are notably higher when compared to SLS and DMLS techniques.

2.1 Printing Parameters

2.1.1 Powder Bed Temperature

To facilitate the sintering procedure, the powder inside the printer requires a heating process [9]. Polymeric materials can be either amorphous or crystalline. For amorphous materials, the powder bed should be set close to their glass transition temperature (T_g). For crystalline materials, the powder bed temperature should be set at least 3–4°C lower than their melting point (M_p). However, most polymers behave as semi-crystalline materials that can exhibit both M_p and T_g. The ideal temperature can be calculated using the following equation:

$$\frac{1}{T} = \frac{w_1}{Mp} + \frac{w_2}{Tg}$$

where W_1 and W_2 represent the weight fractions of each component in the powder mixture [9].

By keeping the powder bed close to either M_p or T_g, less energy would be required from the laser to sinter the object. This would reduce the overall object internal stresses and limit the number of thermal deformations (further described in Sect. 3.4, Liquid-State Melting). For metal and ceramics, powder bed temperatures should be close to their M_p. However, for ceramics the target temperature might be rather high, consequently a lower temperature is usually adopted for the printing process.

2.1.2 Types of Lasers

The distinct differences between PBF printing processes are the type of lasers employed. Ideally, lasers should be selected based on the wavelength absorptance of the materials to be processed. The absorptance is defined as the ratio of the absorbed radiation to the incident radiation [10]. Absorptance depends on the laser wavelength, the nature of the material to be sintered, the shape of the powder particles, the temperature inside the printer and the type of gas in the controlled atmosphere. Generally, a material exhibits a higher absorbance when in its powder form than in dense form. In terms of powder characteristics, the incident radiation is able to penetrate through the particles and further interacts with the underlying particles. This facilitates the energy transmission inside the material [10]. Commercial printers are mainly equipped with either CO_2 (wavelength 10,600 nm) or Nd:YAG laser (wavelength 1064 nm) with a laser power ranging from 50 W to more than 200 W. Table 5.1 shows the different absorption values for metals, ceramics and polymers based for each laser. Generally, Nd:YAG lasers are superior when compared with CO_2 lasers for metals and for carbide ceramics. CO_2 lasers are more efficient for oxide ceramics and polymers.

The higher the absorptance of a powder, the lower the energy required from the laser to sinter. Additives can be used to enhance the powder absorptance. For instance, using a CO_2 laser, the addition of iron to a copper-based powder permits to achieve the same sintering degree with a lower laser power [10]. Energy transmission from the laser can be regulated by tuning laser power as well as the laser scanning speed. Slow laser scanning speeds allow a higher energy transmission.

Recently, more affordable SLS printers were introduced into the market for desktop use. These are equipped with less expensive lasers and are suitable to process polymers powders such as nylon. Fuse 1 [11] and Lisa [12] use an infrared lasers of 10 W, 1066 nm and 5 W, 808 nm respectively. Sintratec Kit [13] use a diode laser of 2.3 W, 445 nm.

Table 5.1 Absorption of powder materials in percentage for Nd:YAG and CO_2 lasers [5, 10]

Powder materials	Nd:YAG (1.06 μm) Absorption	CO_2 (10.6 μm) Absorption
Metals		
Cu	59%	26%
Fe	64%	45%
Sn	66%	23%
Ti	77%	59%
Pb	79%	–
Ceramics		
ZnO	2%	94%
Al_2O_3	3%	96%
SiO_2	4%	96%
SnO	5%	95%
CuO	11%	76%
SiC	78%	66%
Cr_3C_2	81%	70%
TiC	82%	46%
WC	82%	48%
Polymers		
Polytetrafluoroethylene	5%	73%
Polymethylacrylate	6%	75%
Epoxypolyether-based polymers	9%	94%

2.1.3 Particle Size and Shape

Particle size and shape is of the upmost importance in achieving an optimal sintering process and for the fabricated product. Spherical particle sizes ranging from 58 – 180 μm is ideal when employing SLS [14]. This would allow an efficient sintering procedure whereby good flowability is achieved and heat from the laser is homogenously distributed from the surface of the particle to the centre. Particle sizes smaller than the ideal specification (especially less than 10 μm) will reduce overall powder flowability due to the electrostatic forces that provoke aggregation of the particles with increased cohesiveness [15]. Particle sizes that are over 180 μm is also capable of reducing the sintering process. When the laser is in contact with the larger particles, the surface is the only entity that is heated. Due to the uneven transmission of heat, the centre of the particle is often left untreated. Employing larger particles, therefore, would result to an inefficient fusion of the particles and would require more heat from the laser to achieve appropriate sintering. Irregular particle shapes can also deter the sintering process as powder flowability is reduced. Following the specifications of an ideal starting material can result to an effective starting process to SLS.

Fig. 5.3 Scan spacing

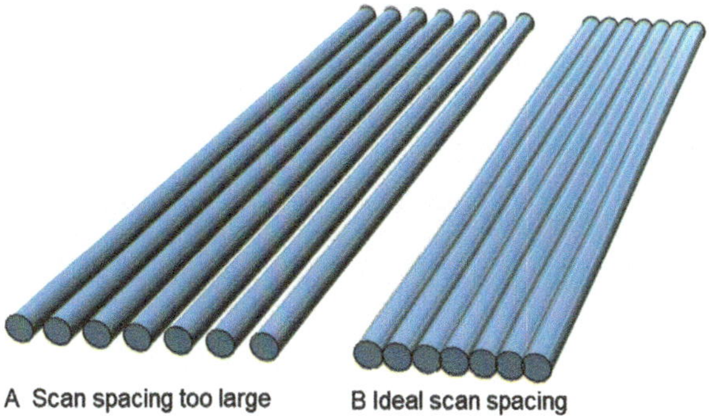

2.1.4 Layer Thickness

Layer thickness usually ranges from 0.1 to 0.3 mm [5]. Thinner layers decrease the surface roughness of the printed object; however, the printing time will be longer. Layer thickness should be also related with the powder particle size. The thickness of a layer should not be thinner than the average particle size of the powder [9].

2.1.5 Additional Parameters

A range of other parameters must be also considered for successful PBF printing. The "scan spacing" is the distance between two parallel scanning lines. If this space is too great, the object cross section could be inefficiently sintered, resulting to weak mechanical properties (Fig. 5.3). Scan spacing has to be set according to the laser beam size [9].

Other parameters include the orientation of the printed object as well as the position on the building platform, the controlled gas atmosphere (inert N_2) and the post-processing procedures. These characteristics are all important to ensure the mechanical integrity of the printed object and for the completion of the printing process [9].

3 Classification

PBF processes can be classified according to the binding mechanism in 4 different groups (Fig. 5.4).

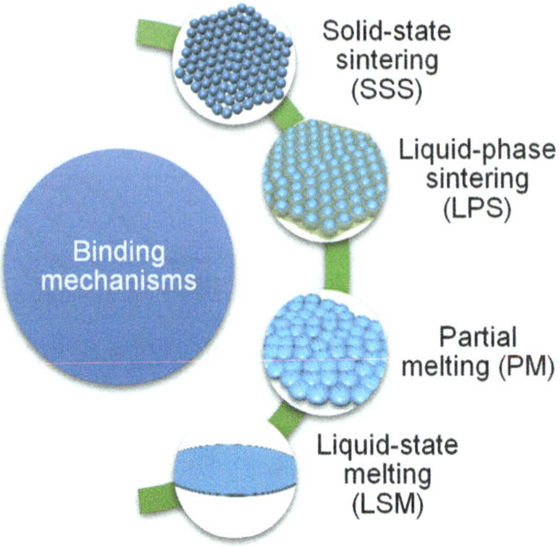

Fig. 5.4 PBF binding mechanisms

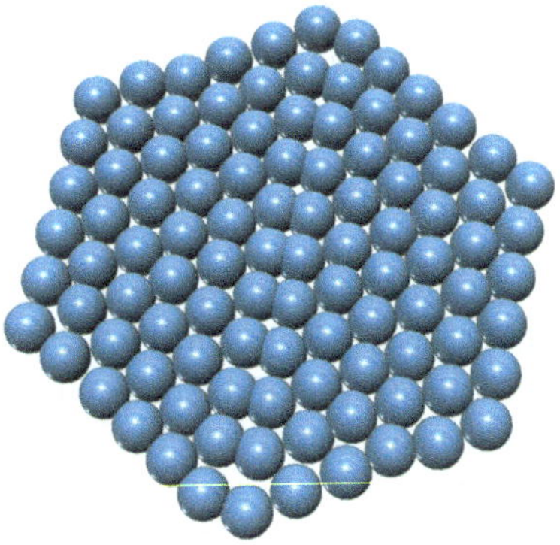

Fig. 5.5 SSS representation. Powder particles are only partially connected

3.1 Solid-State Sintering

Solid-state sintering (SSS) (Fig. 5.5) is a thermal process where the target temperature is set between the melting point (M_p) of the material and $M_p/2$ [16].

The process includes necking (formation of a superficial connection) between adjacent powder particles. Since many empty spaces are left inside the structure, objects manufactured using SSS are always porous. Depending on the sintering parameters, different degrees of porosity can be generated with different levels of pore connectivity. Highly porous objects are lighter and can be produced faster however they can exhibit a lower mechanical strength. SLS and DMLS are essentially two processes that employ SSS. The only difference between SLS and DMLS is the starting material used. SLS process refers to thermoplastic polymers, nylons and ceramics whilst DMLS refers exclusively to metals.

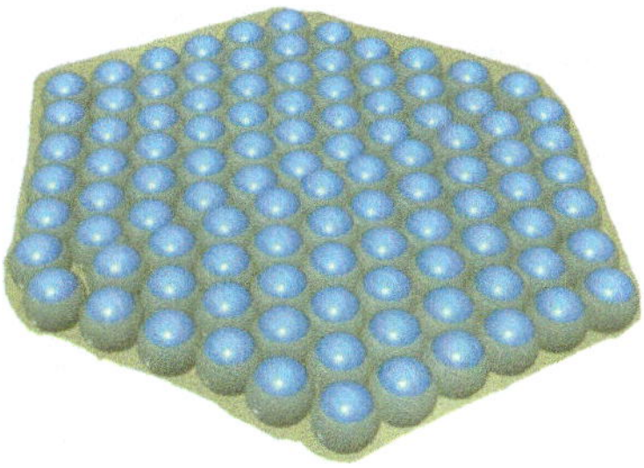

Fig. 5.6 LPS representation. Structural material particles do not sinter or melt. A binder powder melts and infiltrates inside the empty spaces

3.2 Liquid-Phase Sintering

Liquid phase sintering (LPS) (Fig. 5.6) is a thermal process generally used for metals or ceramics with high melting points for the purpose of structural use.

Instead of using high temperatures to sinter or partially melt the structure, a binder material, typically a metal with a lower M_p, is added to the powder mixture. The binder powder particle size is usually much smaller than the structural powder as the higher ratio surface-volume facilitates the melting. The combination of low M_p and small particle size allows the binder powder to melt before the structural powder. The liquid binder is then encouraged to spread between the structural powder particles and thus, fill the empty spaces of the original structure. Remaining spaces (often called green parts) are ascertained after the printing process. This however can be removed with a post-sintering treatment within a furnace through the use of a hot isostatic pressing (HIP) or by the infiltration with low melting point materials [16].

3.3 Partial Melting

In Partial Melting (PM) (Fig. 5.7), powder particles are partially molten and more closely connected when compared with SSS therefore some empty spaces may still remain. However, overall mechanical properties are greater.

3.4 Liquid-State Melting

Liquid-state Melting (LSM) is applied in SLM and requires strong printing conditions. This especially relates for ceramics that use high temperatures and a highly powerful laser, for instance, Zirconia (ZrO_2) melts at 2700 °C. The target temperature is higher than the reported M_p of the starting material to allow the laser to completely melt the powder particles. This would generate a liquid pool of dense material with strong mechanical properties (Fig. 5.8) [17].

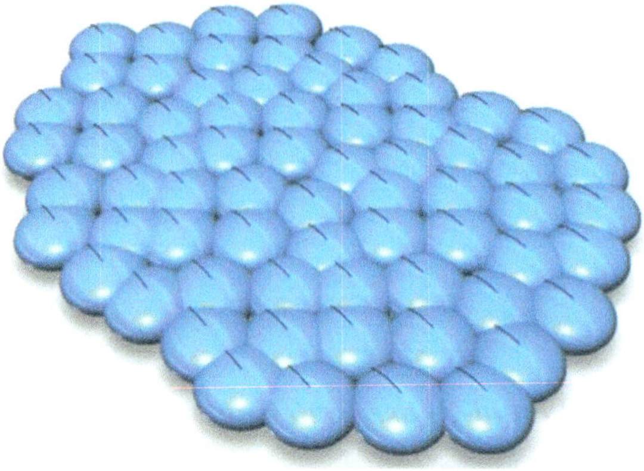

Fig. 5.7 PM representation where powder particles are partially molten

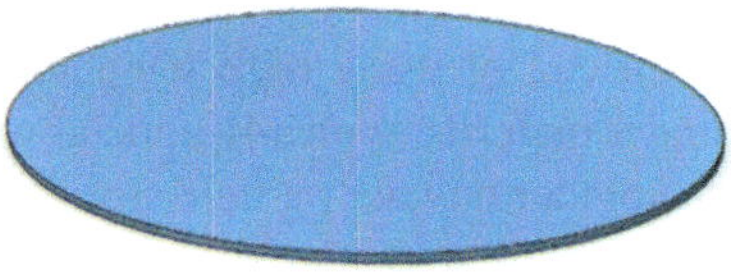

Fig. 5.8 LSM representation. Completely molten particles (liquid pool)

SLM can only be used for a few metals such as stainless steel, titanium, aluminium and cobalt chrome. The limitation in the selection of material is due to the high energy process required resulting to large differences in temperature between the liquid pool and the surrounding powder. Such a difference leads to an increase transmission of heat from the printed object to the surrounding powder, leaving residual stresses inside the object. These stresses may then increase the susceptibility of the final object by provoking cracks or deformations in the designed shape [18]. Additionally, the viscosity and the volume of the material dramatically changes after melting. Liquid droplets tend to minimize their energy by forming spheres (a phenomenon known as "balling"). This further results to instabilities during the printing process [16].

4 Materials in PBF

PBF printing processes can virtually manipulate any powder material or powder blends. In addition, as the allowing of powders are said to be limitless, a wide range of new materials are expected to be introduced to the printing market [19]. In case of materials that do not exhibit a good degree of fusion or sintering, an additional binder can be included in the powder mixture to enhance the sintering process (Fig. 5.6). When the sintering process is complete, the additional binder can be removed by debinding it in a thermal furnace [5].

4.1 Thermoplastic Polymers

The most commonly used polymer is the nylon powder, polyamide (PA) [5]. Different types of PA have been synthesised [20] with PA12 being the most exploited. Several attempts have been made to improve PA12 properties such as blending it with other polymers at different ratios reported in the literature [21, 22]. Other commonly used thermoplastic polymers are polycarbonate (PC) [5], polycaprolactone (PCL) [23–26] and PVA [27].

4.2 Metals

For mechanical purposes within the scope of the aerospace, maritime, automotive industries etc., universally used metals include stainless steel L316, aluminium, titanium Ti64 and cobalt chrome [28]. For medical applications however, metals are required to be biocompatible. Titanium (Ti) alloys such as Ti6Al4V are the most widely used materials for implants, heart valves, bone plates, screws [29] due to their positive corrosion resistance, bio-adhesion and fatigue strength characteristics.

4.3 Ceramics

Ceramics are a class of inorganic and not-metallic materials with outstanding properties. They have a distinctly high mechanical strength, low thermal conductivity and exhibit a resistance to water and corrosive agents. For these reasons, ceramics have been employed to be used in automotive, aerospace and other modern industries [30]. Some ceramics are biocompatible and can be used to manifest body prostheses, tissue engineering scaffolds and dental devices [17]. However, at the current state of technology, the productions of ceramics parts are still challenging due to the high temperatures required for the printing process. The ideal particle size, as well as consolidation mechanisms for porosity reduction, therefore, need to be investigated further [17].

The most frequently used ceramics are ZrO_2 and Silica (SiO_2). ZrO_2 is an oxide ceramic material with outstanding mechanical properties for dental application. In 2007, Bertrand et al. [31] produced a component with 56% density by partial melting (SLS/SLM). In another study, [32], researchers produced a component with almost 100% density containing Al_2O_3/ZrO_2 and a flexural strength of more than 500 MPa through a complete melting of the ceramic powder. However, the final surface roughness of the object was not suitable due to the high pre-heating temperatures required (1600°C). Without high pre-heating temperatures, however, surfaces are smoother, although mechanical strength is compromised due to the formation of micro-cracks.

Silica (SiO_2) is one of the most complex and common ceramic material. It is also a constituent of most glass (up to 60%). In 2003, Tang et al. [33] investigated the production of silica parts and utilised a CO_2 laser. When compared with dense silica, the printed object was reported to have a very poor compressive strength and was therefore unusable. In addition, the article did not disclose data regarding the internal integrity of the object. Hence, cracks may have been present.

5 Industrial Applications of PBF

For industrial application, hard and durable objects are greatly preferred. DMLS and SLM are used in many different fields including the aerospace, maritime and automotive industries (Fig. 5.9) [28, 34].

SLM is mainly used when strong mechanical properties are required as metallic powders are completely molten without retaining any internal porosity [35]. In 2013, NASA [36] described how SLM has been used to manufacture a rocket engine injector. They stated that SLM allowed to print whole objects rather than manufacturing many individual parts that need to be assembled. Using SLM, they assembled the injector by printing just 2 parts instead of 115 parts required if conventional methods were enforced.

A German company, SLM [37], producer of replacement blades for gas turbines, stated that a new blade could be manufactured in just 4 weeks when using SLM when compared with a projected 44-week timeline with the current technologies.

Fig. 5.9 Example of metal object made with DLMS. (Reprinted with permission from [34])

5.1 Medical Applications of PBF

3D printing has been explored for the manufacture of medical devices. In fact, a number of medical devices available in the market already have FDA approval. The FDA has provided a draft guidance for the 3D printing of medical devices [38]. This chapter will go forward by discussing a number of FDA approved medical products and devices, and will cover emerging 3D printing applications in the pharmaceutical field. PBF applications in medicine can be divided in two broad categories: medical metallic devices and tissue engineering.

5.2 Metallic Devices

DMLS and SLM are commonly used to print metallic devices. A wide range of parts today are available to be printed. This includes dental parts [19], trauma medical implants (screws, plates and intramedullary plates, intramedullary rods) (Fig. 5.9) and orthopaedic medical devices (knee joint devices and hip joint devices) (Fig. 5.10) [29].

Bio-adhesion is an important requirement for medical devices to assure the integration of the device through the bone and for bone tissue regeneration. Comparative investigations between steel and titanium implants show that bone regeneration with steel devices is much lower than titanium devices. In particular, Ti6Al4V showed excellent biochemical effect in bone growth, as well as positive biomechanical properties [29].

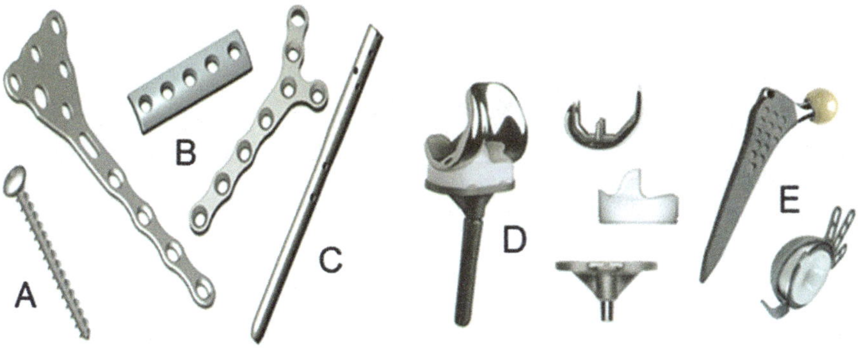

Fig. 5.10 Trauma medical implants: screws (**a**), plates and intramedullary plates (**b**), intramedullary rods (**c**). Orthopaedic medical devices: knee joint devices (**d**) and hip joint devices (**e**). (Reprinted with permission from [29])

5.3 Tissue Engineering

SLS has been widely recognised to have a great potential in tissue engineering (TE), a new field firstly introduced in the mid-1980s. TE applies engineering innovations to life science with the aim of either repairing or creating human tissues such as bone, cartilage and ligaments. TE methods include the presence of a 3D bio-resorbable scaffold (Fig. 5.11) to guide the cells attachment and to support their differentiation and proliferation. Unfortunately, lack of suitable biomaterials and manufacturing methods have delayed the progression of this area [40].

The first application of SLS in TE was in 1996 when Linda and Michael Cima [41] announced that their first patent was able to manufacture medical devices for the controlled release of active agents or medical devices containing cells with the

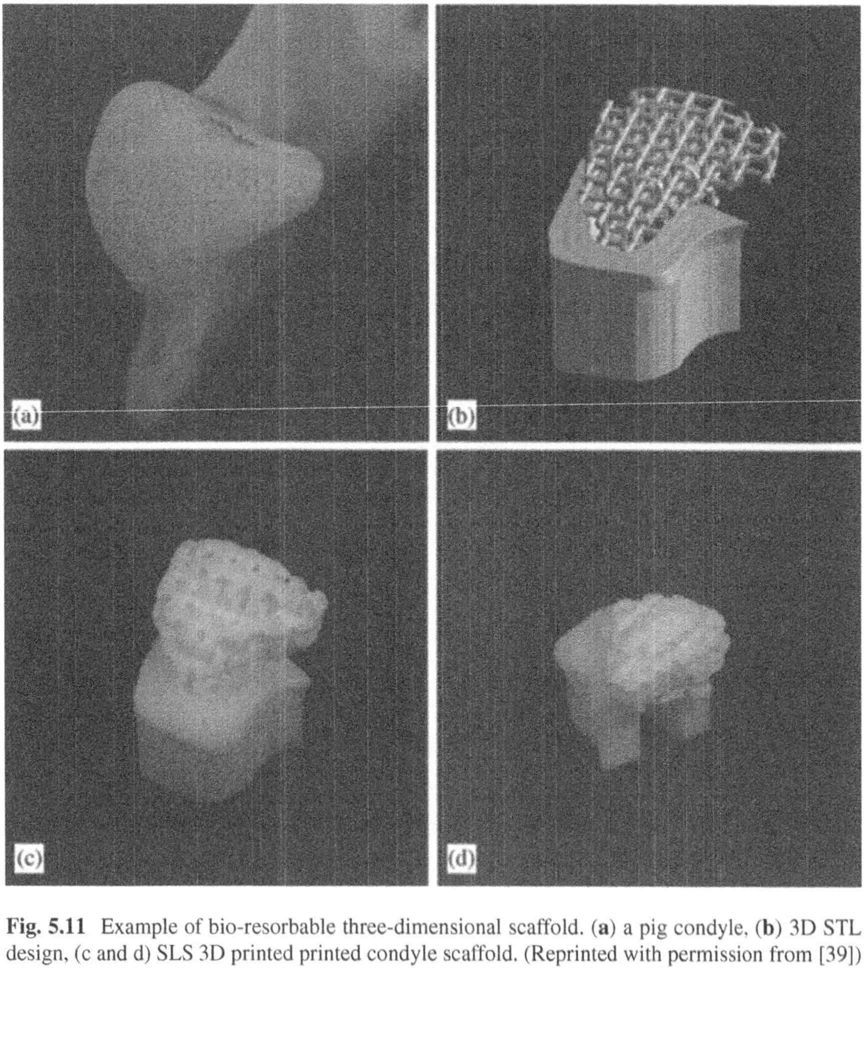

Fig. 5.11 Example of bio-resorbable three-dimensional scaffold. (**a**) a pig condyle, (**b**) 3D STL design, (c and d) SLS 3D printed printed condyle scaffold. (Reprinted with permission from [39])

use of SLS. In 1997, SLS was coupled with computed tomography (CT) to scan two skulls (one from a child and one from an adult) and a femur [42]. An SLS Sinterstation 2000 (DMT Corporation, Austin, TX) was employed whilst nylon was chosen as the structural powder material. The final printed product dimensions were in good agreement with the CT scan data, exhibiting less than 1 mm deformation in each direction.

In 2003, Tan et al. [43] investigated the use of SLS in tissue engineering with the aim of creating a temporary 3D scaffold, essential for the proliferation of the tissue or organ growth. Polyether ether ketone (PEEK) and hydroxyapatite (HA) powders were physically blended and used to produce the scaffold. The advantage of using SLS is due to its solvent free process and the possibility of controlling the porosity formation, fundamental to encourage tissue growth. In 2004, Chua et al. [27] used a bio-composite blend of PVA and HA as a powder material to fabricate scaffolds for craniofacial and joint defects using SLS. PVA has been selected due to its similar tensile strength to human articular cartilages, the ability to form complex shapes and its good adhesion properties. HA is a bioactive material with osteoconductivity properties. HA can create bonds to bones with a strength similar to that of a 3–6-months year old and it is also capable of increasing the production of collagen. Tests on the bioactivity of HA showed that the laser sintering process did not provoke any alteration and that HA was still bioactive in the environment. Therefore, suitable scaffolds can be produced using a mixture of PVA and HA. In 2005, Tan et al. [44] screened several biocompatible polymers from PVA, PEEK, Polycaprolactone (PCL) to poly-L-lactic acid (PLLA) with the addition of HA. They demonstrated the suitability of those polymers to produce TE scaffolds for the regeneration of tissues. Scaffolds can be very porous depending on the printing parameters.

In 2005, Williams et al. [39] from University of Michigan used PCL to design scaffolds for bone tissue engineering. A good degree of porosity was achieved (up to 55%), however, the designed porosity was much higher (about 80%). Additionally, some porosity (20%) was undesirably present in the solid areas. These issues were due to an inappropriate particle size that was trapped in the pores. This presents one of the limitation in SLS; when the particle size is too large, a reduction of the object resolution is achieved in conjunction with an increased granularity localised to the edges. On the contrary, too small particles sizes have poor flowability and are difficult to remove from inside the pores after the printing process. To improve the overall structure of those scaffolds, the same research group [40] further investigated the properties of PCL scaffolds by improving the printing parameters using a 2^5 full factorial study design. A more accurate selection of the particle size, together with an optimization of the printing parameters, allowed to produce scaffolds with 94% density in the solid regions. This resulted to a more regular pore geometry of the porous regions which was within 8% of the designed dimensions.

PCL scaffolds were further investigated in 2010 with the inclusion of 30% of HA [45]. Printed scaffolds, however, resulted in larger dimensions when compared with the designed ones. Additionally, different mechanical behaviours were obtained based on the manufacturing direction.

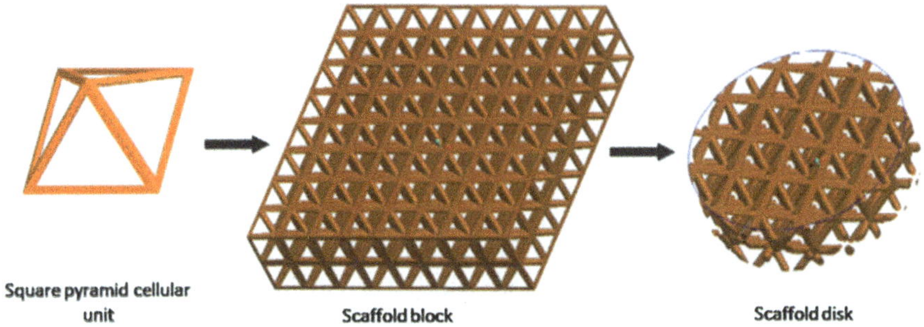

Fig. 5.12 3D design of the flexible scaffold disk. (Reprinted with permission from [47])

In 2010, Yeong et al. [46] investigated the opportunity to prepare PCL scaffolds for cardiac tissue engineering designed with an automated algorithm named "computer-aided system for tissue scaffolds" (CASTS). The cellular unit was a square pyramid (Fig. 5.12). This peculiar structure allowed to print a scaffold with a low stiffness (345 kPa) and 85% porosity that could be used for the regeneration of the cardiac tissue. Myoblast cells were cultured *in vitro* and after 21 days, fluorescence images showed the cells distributed throughout the scaffold.

In 2010, Duan et al. [47, 48] fabricated 3D scaffolds for bone tissue engineering. They used calcium phosphate (Ca-P)/poly(hydroxybutyrate–co-hydroxyvalerate) PHBV and carbonated hydroxyapatite (CHAp)/poly(L-lactic acid) (PLLA) nanocomposite microspheres. Future works include resolutions below 400-500 μm and more spherical powder particle sizes that would be easily removable with traditional methods or pressured air [49].

5.4 *PBF in Pharmaceutics*

Unlike for medical devices and implants, no FDA guidance has been provided for the 3D printing of solid oral dosage forms.

DMLS and SLM are process that incorporate strong lasers, SLS is the only appropriate method that could be used for pharmaceutical application. The first attempt to apply SLS in pharmaceutics was in 2001 from Nanyang Technological University in Singapore. Leong et al. [50] used fine nylon (average 20μm) from DTM Corporation to print 8 × 8 × 8mm cubes with an SLS printer using a CO_2 laser. The aim of the study was to investigate how the printing parameters would have influenced the overall object porosity. Cubes were then infiltrated with a model drug, methylene blue dye, followed by a visual inspection of the dye penetration into the matrices. It was clear that by adjusting the printing parameters, such as laser scanning speed and laser power, different controlled release profiles could be obtained. From this study, it was also identified that cubes had two dense walls due to the dwell time at the beginning and end of each layer scan (Fig. 5.13).

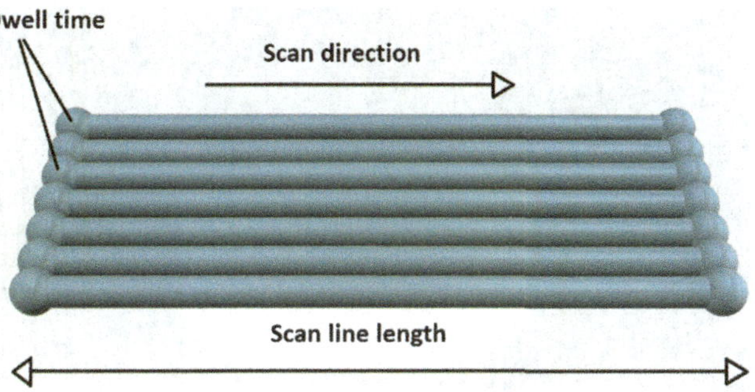

Fig. 5.13 Formation of dense walls due to the dwell time of the laser

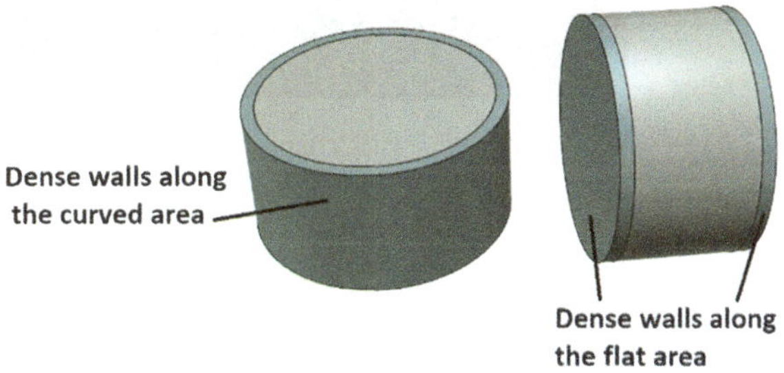

Fig. 5.14 Location of dense walls based on the different printing orientation

The material in dense walls is partially molten (PM) whilst the middle of the structure is sintered. Dense walls, being stronger and less porous, were capable of decreasing the rate of diffusion of the drug from the internal matrix.

A second study from Leong [51], aimed at understanding the formation of the dense walls based on the printing orientation (Fig. 5.14). Duraform from DMT Corporation (Polyamide, PA) was used.

The aim was also to understand the role of the powder bed temperature. It was observed a correlation of the increasing powder bed temperature with the reduced porosity.

In a following study [52], the same group from Nanyang Technological University investigated the role of laser scanning speed, varying pore density, interconnectivity and the formation of dense walls evaluating the release of a dye. Duraform was used as the starting powder material and methylene blue dye was selected as the model drug. The drug was infiltrated into the porous matrix after the printing process. Release tests were performed to evaluate the variations of the final structure according to the different parameters investigated. From the results, a strong correlation between printing parameters and release profile was present, indicating the possibility of achieving catered drug release profile by adjusting printing parameters. In 2006, Leong et al. [14] investigated the possibility to build drug delivery devices (DDD) using either PCL or PLLA. Seven different laser powers (from 1 W to 7 W) were tested whilst other parameters were kept constant. The same method was applied to 6 different laser scanning speeds (from 1270 mm/s to

Fig. 5.15 Image of a cylindrical drug delivery device printed using SLS with PCL. Settings were 3 W (inner region)–4 W (outer region), scan speed 5080 mm/s (200in/s) and part bed temperature 40 °C. (Reprinted with permission from [14])

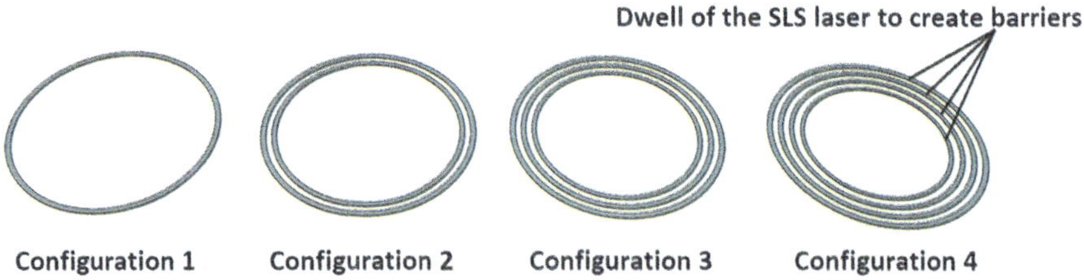

Fig. 5.16 Representation of 4 different configurations made by creating molten barriers to slow down the release. Barriers can be made using a dwell of the laser

7620 mm/s), keeping the other parameters constant. The porosity of all the products were measured and a clear correlation was achieved. By either reducing the laser power or increasing the laser speed, a higher porosity level was obtained. Optimal settings (Fig. 5.15) were then gathered from a balance of present porosity (of at least 50%) and sintering connections.

For the first time, in 2007, [24] researchers were successfully able to sinter a powder obtained from a mixture of polymer and drug. PCL was chosen as polymer while the model drug was methylene blue dye. Results showed that drug distribution of methylene blue dye was uniform in the device. Dissolution tests showed a burst release followed by a constant release of the drug. To limit the burst release, a series of barrier rings (from 1 to 4) were created by introducing a dwell of the laser (Fig. 5.16). The best reduction in the burst release was obtained from the DDD with 4 barriers.

Table 5.2 Printlets composition [7]

Formulation[a]	Kollicoat IR (%)	Eudragit L100-55 (%)	Paracetamol (%)
K5	92	–	5
K20	77	–	20
K35	62	–	35
E5	–	92	5
E20	–	77	20
E35	–	62	35

[a]All formulations contain 3% w/w Candurin® Gold Sheen

Work on SLS to produce DDD was progressed forward from the Federal University of Santa Catarina in Brazil. Salmoria et al. [25] used SLS technology with a CO_2 laser to print a DDD based on PCL containing progesterone. The drug was included in the powder mixture at 15% w/w before printing. Devices were shown to release the drug of up to 14% in 80 days. Drug release was therefore either not complete or some degradation of the drug took place, as suggested in a more recent study [53]. However, no investigation regarding the effect of the laser to the drug degradation were reported. In a following study, Salmoria et al. [54] improved the previous device by investigating the effects of the particle size and the laser energy. They produced 4 different DDD that released progesterone from a PCL matrix of up to 45% in 80 days. Again, no studies regarding a possible laser degradation of the drug were reported. In 2016, the same group [55] investigated the production of PCL containing 15% w/w Ibuprofen implants for orbital floor repair. They used an SLS printer equipped with a CO_2 laser to produce the implants. Dissolution results showed the drug to be release within 10 h. However, it was not clearly stated which percentage of drug was released and if laser drug degradation occurred. In 2017, the same research group [23, 56] demonstrated with two further studies that the feasibility of producing implants using either Polyethylene (PE) or PCL containing Fluorouracil.

In 2017, a research group from UCL (University College London) manufactured the first oral product using SLS technology [7]. For the first time, two pharmaceutical polymers commonly used in the pharmaceutical industry, Eudragit L100-55 and Kollicoat IR with prolonged and immediate release properties respectively, were employed to produce 3D printed tablets (herein known as printlets: a term that we have coined to describe 3DP oral dosage forms) containing an active ingredient (paracetamol) [56]. The drug was included in the powder mixtures before starting the printing process in three different drug loadings for each polymer (Table 5.2).

Printing parameters (powder bed temperature 110 °C, laser power 2.3 W and laser scanning speed 90 mm/s) were consistent for printing all the formulations. The printer employed in this study was a Sintratec Kit (AG, Brugg, Switzerland) working with a diode laser of 445 nm at 2.3 W. To enhance the powder absorptance, 3% of a pharmaceutical grade additive (Candurin® Gold Sheen) was included in all the formulations. Printlets were successfully printed with a good degree of reproducibility (Fig. 5.17).

Fig. 5.17 Image of the printlets, on the top from left to right K5, K20, K35; on the bottom, from left to right E5, E20, E35. (Reprinted with permission from [7])

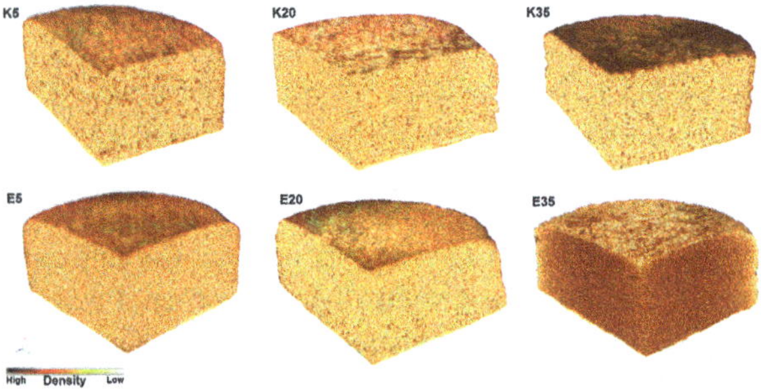

Formulation	Closed porosity (%)	Open porosity (%)
K5	0.2	33.3
K20	0.4	24.6
K35	0.8	24.0
E5	0.1	29.3
E20	1.0	20.3
E35	4.7	1.0

Fig. 5.18 Left: X-Ray micro-CT images of a quarter section of the printlets showing different densities. on the top from left to right K5, K20, K35; on the bottom, from left to right E5, E20, E35. Right: Porosity values of the printlets. (Reprinted with permission from [7])

X-Ray micro-CT was employed to visualize the internal structure and to calculate the porosity of the printlets (Fig. 5.18).

Figure 5.19 shows SEM images of the printlets (vertical section). It is evident that by increasing the drug content, the Printlets resulted from being sintered to partially molten due to the plasticizing effect of the drug which reduced the T_g of the polymers.

Since the laser degradation of the drug was a great concern, degradation studies were carried out and no drug degradation was observed after the printing process. Drug dissolution profiles showed a pH-independent profile for Kollicoat IR formulations (Fig. 5.20). The difference in drug release was due to the overall energy intake during the printing process. Printlets containing 5% drug were sintered and there was a high pore connectivity leading to a faster release (2 h).

Fig. 5.19 SEM images of the printlets vertical sections, on the top from left to right K5, K20, K35. On the bottom from left to right E5, E20, E35. (Reprinted with permission from [7])

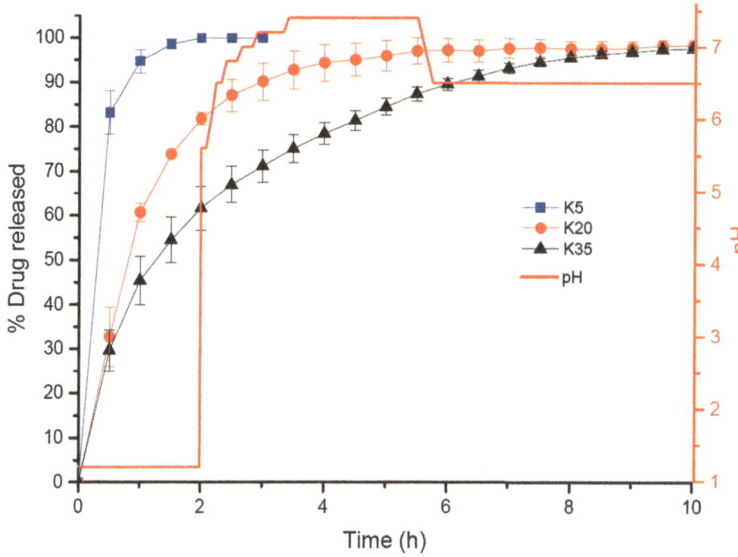

Fig. 5.20 Drug dissolution profiles from Kollicoat printlets. The red line shows the pH values of the media. (Reprinted with permission from [7])

Printlets formulated with 35% drug were still molten with high porosity, but exhibited low pore connectivity, and thus, led to a longer dissolution time (8 h).

Release profile for Eudragit formulations (Fig. 5.21) showed an enteric release with limited drug release (less than 20%) in the acidic environment during the first 2 h, followed by an increased release in intestinal conditions. More recent work has shown that SLS 3D printing is feasible to fabricate printlets with accelerated drug

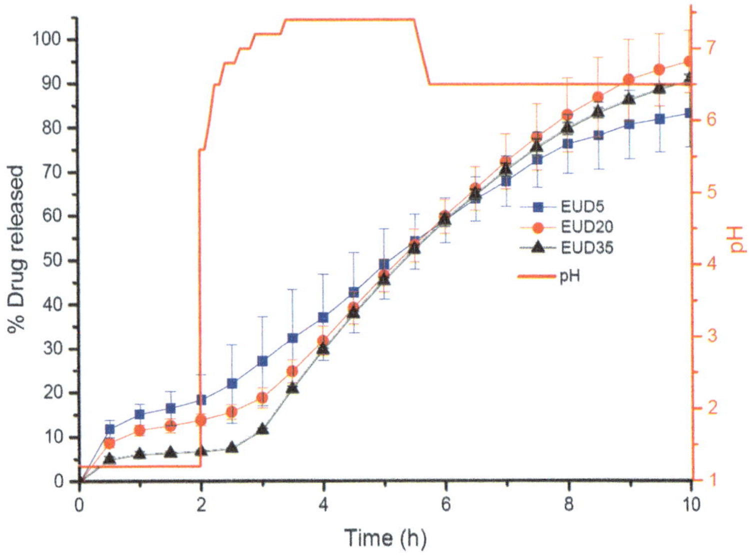

Fig. 5.21 Drug dissolution profiles from Eudragit printlets. The red line shows the pH values of the media. (Reprinted with permission from [7])

release, orally disintegrating properties [57] and complex dimensions such as a gyroid lattice structure [58]. This investigation has confirmed that SLS is amenable to the pharmaceutical research of modern medicine manufacture [56, 59–61].

6 Conclusion

From the invention of the first prototypes almost 30 years ago, PBF technologies have finally gained traction and started to revolutionize the whole manufacturing process. Fast production of spare parts in the aerospace, maritime and automotive industries can now be done in just one step, rather than assembling a multitude of single parts that were manufactured with different processes. Personalised medical devices including dental parts, trauma medical implants and orthopaedic medical devices will be just as easy to manufacture, less expensive and tailored specifically for the patient. Applications in tissue engineering are very promising for the regeneration of tissues and may even be soon applied in clinics. Interestingly, the pharmaceutical field may soon be subject to disruption by virtue of innovation due to a very recent manufacture of the first pharmaceutical oral product by SLS in 2017. The recent commercialization of many new printer models for industrial use, as well as for desktop use, will expand the potential application of PBF and ultimately, uncover a very promising future for 3D printing.

References

1. Shellabear M, Nyrhilä O. DMLS – Development history and state of the art. 2004.
2. Beaman JJ, Deckard CR. Selective laser sintering with assisted powder handling. US 4938816 A. 1990.
3. Geiger M, Vollertsen F. Laser assisted net shape engineering. Proceedings of the LANE '94 conference, Erlangen, Oct 12–14, 1994.
4. Meiners, Wissenbach, Gasser. Selective laser sintering at melting temperature. German patent DE 19649865, filed December 2nd, 1996, Published Feb 12, 1998. 1996.
5. Kruth JP, Wang X, TL FL. Lasers and materials in selective laser sintering. Assem Autom. 2003;23(4):357–71.
6. Louvis E, Fox P, Sutcliffe CJ. Selective laser melting of aluminium components. J Mater Process Technol. 2011;211(2):275–84.
7. Fina F, Goyanes A, Gaisford S, Basit AW. Selective laser sintering (SLS) 3D printing of medicines. Int J Pharm. 2017;529(1):285–93.
8. Leong KF, Cheah CM, Chua CK. Solid freeform fabrication of three-dimensional scaffolds for engineering replacement tissues and organs. Biomaterials. 2003;24(13):2363–78.
9. Gibson I, Shi D. Material properties and fabrication parameters in selective laser sintering process. Rapid Prototyp J. 1997;3(4):129–36.
10. Tolochko NK, Laoui T, Khlopkov YV, Mozzaharov SE, Titov VI, Ignatiev MB. Absorptance of powder materials suitable for laser sintering. Rapid Prototyp J. 2000;6(3):155–61.
11. Formlabs. Fuse 1 Tech Specs. https://formlabs.com/3d-printers/fuse-1-tech-specs/. 2017.
12. Sinterit. Sinterit Lisa. http://wwwsinterit.com/?gclid=EAIaIQobChMI5vOpvZbU1gIVB40bCh2CzQCUEAAYAiAAEgLLtvD_BwE. 2017.
13. Sintratec. Sintratec Kit SLS printer. http://sintratec.com/products/kit. 2017.
14. Leong KF, Chua CK, Gui WS. Verani. Building porous biopolymeric microstructures for controlled drug delivery devices using selective laser sintering. Int J Adv Manuf Technol. 2006;31(5–6):483–9.
15. Schulze D. Powders and bulk solids: behavior, characterization, storage and flow. Berlin Heidelberg New York: Springer; 2008.
16. Kruth J-P, Mercelis P, Van Vaerenbergh J, Froyen L, Rombouts M. Binding mechanisms in selective laser sintering and selective laser melting. Rapid Prototyp J. 2005;11(1):26–36.
17. Sing SL, Yeong WY. Direct selective laser sintering and melting of ceramics: a review. Rapid Prototyp J. 2017;23(3):611–23.
18. Mercelis P, Kruth J-P. Residual stresses in selective laser sintering and selective laser melting. Rapid Prototyp J. 2006;12(5):254–65.
19. Gebhardt A, Schmidt F-M, Hötter J-S, Sokalla W, Sokalla P. Additive manufacturing by selective laser melting the realizer desktop machine and its application for the dental industry. Phys Procedia. 2010;5(Part B):543–9.
20. Stratasys. Selective laser sintering materials. https://wwwstratasysdirect.com/materials/laser-sintering/. 2017.
21. Salmoria GV, Lauth VR, Cardenuto MR, Magnago RF. Characterization of PA12/PBT specimens prepared by selective laser sintering. Opt Laser Technol. 2018;98:92–6.
22. Salmoria GV, Leite JL, Paggi RA. The microstructural characterization of PA6/PA12 blend specimens fabricated by selective laser sintering. Polym Test. 2009;28(7):746–51.
23. Salmoria GV, Vieira FE, Ghizoni GB, Marques MS, Kanis LA. 3D Printing of PCLFluorouracil tablets by selective laser. 2017.
24. Leong KF, Wiria FE, Chua CK, Li SH. Characterization of a poly-epsilon-caprolactone polymeric drug delivery device built by selective laser sintering. Biomed Mater Eng. 2007;17(3):147–57.
25. Salmoria GV, Klauss P, Zepon K, Kanis LA, Roesler CRM, Vieira LF. Development of functionally-graded reservoir of PCL/PG by selective laser sintering for drug delivery devices. Virtual Phys Prototyp. 2012;7(2):107–15.

26. Salmoria GV, Klauss P, Roesler CRM, Kanis LA. Structure and mechanical properties of PCL/PG devices prepared by selective laser sintering for drug delivery applications. 2013 (55607):V01AT20A016.
27. Chua CK, Leong KF, Tan KH, Wiria FE, Cheah CM. Development of tissue scaffolds using selective laser sintering of polyvinyl alcohol/hydroxyapatite biocomposite for craniofacial and joint defects. J Mater Sci Mater Med. 2004;15(10):1113–21.
28. Stratasys. Direct metal laser sintering materials. https://wwwstratasysdirect.com/materials/direct-metal-laser-sintering/. 2017.
29. Novakov T, Jackson MJ, Robinson GM, Ahmed W, Phoenix DA. Laser sintering of metallic medical materials—a review. Int J Adv Manuf Technol. 2017;93:2723.
30. Dutta S. Fracture toughness and reliability in high-temperature structural ceramics and composites: prospects and challenges for the 21st century. Bull Mater Sci. 2001;24(2):117–20.
31. Bertrand P, Bayle F, Combe C, Goeuriot P, Smurov I. Ceramic components manufacturing by selective laser sintering. Appl Surf Sci. 2007;254(4):989–92.
32. Yves-Christian H, Jan W, Wilhelm M, Konrad W, Reinhart P. Net shaped high performance oxide ceramic parts by selective laser melting. Phys Procedia. 2010;5(Part B):587–94.
33. Tang Y, Fuh JYH, Loh HT, Wong YS, Lu L. Direct laser sintering of a silica sand. Mater Des. 2003;24(8):623–9.
34. 3DSystems. Metal printers. https://uk3dsystems.com/3d-printers#metal-printers. 2017.
35. Abe F, Osakada K, Shiomi M, Uematsu K, Matsumoto M. The manufacturing of hard tools from metallic powders by selective laser melting. J Mater Process Technol. 2001;111(1):210–3.
36. NASA. NASA Tests Limits of 3-D Printing with Powerful Rocket Engine Check. https://www.nasa.gov/press/2013/august/nasa-tests-limits-of-3-d-printing-with-powerful-rocket-engine-check/#WcjOmciGNPZ. 2013.
37. SLM. Nickel based alloys Turbine blade. https://slm-solutions.com/products/accessories-and-consumables/slm-metal-powder. 2016.
38. FDA. Technical considerations for additive manufactured devices: food and drug administration; 2017. Available from: https://www.fda.gov/downloads/MedicalDevices/DeviceRegulationandGuidance/GuidanceDocuments/UCM499809.pdf.
39. Williams JM, Adewunmi A, Schek RM, Flanagan CL, Krebsbach PH, Feinberg SE, et al. Bone tissue engineering using polycaprolactone scaffolds fabricated via selective laser sintering. Biomaterials. 2005;26(23):4817–27.
40. Partee B, Hollister SJ, Das S. Selective laser sintering process optimization for layered manufacturing of CAPA[sup ®] 6501 Polycaprolactone bone tissue engineering scaffolds. J Manuf Sci Eng. 2006;128(2):531.
41. Cima LG, Cima MJ. Patent preparation of medical devices by solid free-form fabrication methods. 1996.
42. Berry E, Brown JM, Connell M, Craven CM, Efford ND, Radjenovic A, et al. Preliminary experience with medical applications of rapid prototyping by selective laser sintering. Med Eng Phys. 1997;19(1):90–6.
43. Tan KH, Chua CK, Leong KF, Cheah CM, Cheang P, Abu Bakar MS, et al. Scaffold development using selective laser sintering of polyetheretherketone–hydroxyapatite biocomposite blends. Biomaterials. 2003;24(18):3115–23.
44. Tan KH, Chua KC, Leong KF, Cheah CM, Gui WS, Tan WS, Wiria FE. Selective laser sintering of biocompatible polymers for applications in tissue engineering. Biomed Mater Eng. 2005;15(1):11–124.
45. Eosoly S, Brabazon D, Lohfeld S, Looney L. Selective laser sintering of hydroxyapatite/poly-ε-caprolactone scaffolds. Acta Biomater. 2010;6(7):2511–7.
46. Yeong WY, Sudarmadji N, Yu HY, Chua CK, Leong KF, Venkatraman SS, et al. Porous polycaprolactone scaffold for cardiac tissue engineering fabricated by selective laser sintering. Acta Biomater. 2010;6(6):2028–34.
47. Duan B, Wang M, Zhou WY, Cheung WL, Li ZY, Lu WW. Three-dimensional nanocomposite scaffolds fabricated via selective laser sintering for bone tissue engineering. Acta Biomater. 2010;6(12):4495–505.

48. Duan B, Cheung WL, Wang M. Optimized fabrication of Ca-P/PHBV nanocomposite scaffolds via selective laser sintering for bone tissue engineering. Biofabrication. 2011;3(1):015001.
49. Chia HN, Wu BM. Recent advances in 3D printing of biomaterials. J Biol Eng. 2015;9(1):4.
50. Leong KF, Phua KKS, Chua CK, Du ZH, Teo KOM. Fabrication of porous polymeric matrix drug delivery devices using the selective laser sintering technique. Proc Inst Mech Eng H J Eng Med. 2001;215(2):191–2.
51. Leong K. Characterization of SLS parts for drug delivery devices. Rapid Prototyp J. 2001;7(5):262–8.
52. Cheah CM, Leong KF, Chua CK, Low KH, Quek HS. Characterization of microfeatures in selective laser sintered drug delivery devices. Proc Inst Mech Eng H J Eng Med. 2002;216(6):369–83.
53. Alhnan MA, Okwuosa TC, Sadia M, Wan K-W, Ahmed W, Arafat B. Emergence of 3D printed dosage forms: opportunities and challenges. Pharm Res. 2016;33(8):1817–32.
54. Salmoria GV, Klauss P, Zepon KM, Kanis LA. The effects of laser energy density and particle size in the selective laser sintering of polycaprolactone/progesterone specimens: morphology and drug release. Int J Adv Manuf Technol. 2012;66(5–8):1113–8.
55. Salmoria GV, Cardenuto MR, Roesler CRM, Zepon KM, Kanis LA. PCL/ibuprofen implants fabricated by selective laser sintering for orbital repair. Procedia CIRP. 2016;49:188–92.
56. Trenfield SJ, Awad A, Goyanes A, Gaisford S, Basit AW. 3D printing pharmaceuticals: drug development to frontline care. Trends Pharmacol Sci. 2018;39(5):440–51.
57. Fina F, Madla CM, Goyanes A, Zhang J, Gaisford S, Basit AW. Fabricating 3D printed orally disintegrating printlets using selective laser sintering. Int J Pharm. 2018;541(1-2):101–7.
58. Fina F, Goyanes A, Madla CM, Awad A, Trenfield SJ, Kuek JM, Patel P, Gaisford S, Basit AW. 3D printing of drug-loaded gyroid lattices using selective laser sintering. Int J Pharm. 2018; https://doi.org/10.1016/j.ijpharm.2018.05.044.
59. Lepowsky E, Tasoglu S. 3D printing for drug manufacturing: a perspective on the future of pharmaceuticals. Int J Bioprint. 2018;4(1):119.
60. Awad A, Trenfield SJ, Gaisford S, Basit AW. 3D printed medicines: A new branch of digital healthcare. 2018;548(1):586–596.
61. Awad A, Trenfield SJ, Goyanes A, Gaisford S, Basit AW. Reshaping drug development using 3D printing. Drug Discov Today. 2018; https://doi.org/10.1016/j.drudis.2018.05.025.

Chapter 6
Fused Deposition Modelling: Advances in Engineering and Medicine

Atheer Awad, Simon Gaisford, and Abdul W. Basit

Abstract Fused deposition modelling (FDM) is a material extrusion process that utilises thermoplastic polymers to construct objects in an additive manner. This singular platform offers new prospects for the fields of engineering and medicine, making it a widely applicable 3D printing technology. As such, FDM has multitude applications ranging from prototyping and consumer products to industrial processing. Known for its low cost, diversity, portability and simplicity, this technology has the potential to revolutionise the field of medicine by providing means to introduce innovative concepts, including dual printing and contemporaneous forms of drug delivery systems. Its ability to precisely design and fabricate novel structures allows more control over drug performance, enhancing the quality of treatment and avoiding constraints often imposed by conventional manufacturing processes. This chapter overviews the current industrial advancements of the FDM technology in engineering and medicine, discussing the potential advantages of implementing this technique in pharmaceutical production and providing an insight on its current limitations.

Keywords Three-dimensional printing · Fused deposition modeling · Personalized healthcare · Fused filament fabrication · Digital pharmacy · Drug delivery systems

---

A. Awad (✉)
Department of Pharmaceutics, UCL School of Pharmacy,
University College London, London, UK
e-mail: atheer.awad.15@ucl.ac.uk

S. Gaisford · A. W. Basit
Department of Pharmaceutics, UCL School of Pharmacy,
University College London, London, UK

FabRx Ltd., Ashford, Kent, UK

A. W. Basit, S. Gaisford (eds.), *3D Printing of Pharmaceuticals*,
AAPS Advances in the Pharmaceutical Sciences Series 31,
https://doi.org/10.1007/978-3-319-90755-0_6

1 Introduction

Fused deposition modelling (FDM), also known as fused filament fabrication (FFF), is an additive manufacturing technology that utilises a 3D design to create an object by depositing molten materials on a platform in a layer-by-layer manner. This technology was first developed by Scott Crump in 1988 [1]. The idea started when Crump was attempting to create a toy for his daughter using a simple glue gun. He replaced the conventional glue stick with a fusion of polyethylene and candle wax and used it to construct the toy layers one over the other. Consequently, this led to the idea of developing an automated version of the process. In 1989, Crump and his wife patented the technology and co-founded their own company, Stratasys, to commercialise their product [2, 3]. Due the expiration of the patent in 2009, there was a significant increase in the availability of FDM printers. This remarkably reduced the prices of printers by almost 10-fold, making them more available to consumers. Nowadays, the FDM technology is the most widely used 3D printing technology.

Typically, filaments made out of thermoplastic materials have been used as the feedstock of FDM printers. Examples of thermoplastics include polylactic acid (PLA), acrylonitrile butadiene styrene (ABS), polyvinyl alcohol (PVA) and polycarbonate (PC) and are filaments commercially available as spools with different colours. The type of printing material will greatly influence the final 3D printed object in terms of mechanical strength and durability. This is mainly related to the different properties and melting temperatures associated with each type of polymer. Therefore, the selection of the appropriate type of material can be critical to achieve a successful printing process.

Today, the filament production technology has advanced, allowing the incorporation of different materials, such as wood, metal, nylon and cork fills, which can be added into the thermoplastic blend. This gives the 3D printed objects a more natural appearance, taking us away from the conventional plastic view that most people are used to. For instance, the laywoo-d3 filament, which is composed of a combination of recycled wood and polymer, possesses the smell and texture of wood. Thus, once grinded, sanded and painted, the final printed objects can be deemed as true wooden sculptures. Likewise, the addition of a colour-diode or glowing materials, which respond to a certain stimulus, opens the gates to a whole new world known as 4D printing (For more details, see Chap. 8). In addition, the invention of the carbon-infused filaments that have electrically-conductive properties, allows the printing of fully functional electronic devices. Thus, the possibilities of this technology can be considered endless.

Whilst FDM only deals with thermoplastic filaments, it constitutes a small branch of a wider technology known as "material extrusion". Material extrusion expands to include materials such as clays, waxes, gels and pastes. With the latter two being in the form of semi-solids, they require the use of syringes to be fed into the 3D printers. This form of material extrusion is classified as semi-solid extrusion and will be further discussed in more details in Chap. 7.

2 The FDM Process

The process consists of designing a 3D shape using a CAD software. This shape is then loaded into the software of the printer, usually in the form of a stereolithography (.stl) file. A thermoplastic filament is then fed into the printer. This filament is melted through a metal nozzle under controlled temperatures. The molten thread-like material is subsequently deposited on the flat base of the printer known as the platform. The movement of the nozzle in an x-y direction helps in creating the desired 3D object. The printer starts off by printing the exterior lining of the object, known as the shell, then fills it based on the pre-determined infill percentage. As the filament cools down, it attaches to the previous layer. Upon the completion of each layer, the platform is lowered to allow enough space for a new layer to be printed. The process is repeated until the object is completed (Fig. 6.1).

Some objects have complex structures that makes their printing process difficult. For instance, round objects lack a flat surface, making it challenging for such objects to adhere to the surface of the platform. Thus, a form of an additive material, to aid in holding the structures in their positions, is needed. This is known as a "support material" [4]. At the end of the printing process, the support material is detached wherein based on the form of detachment, support materials can be classified into two types, the soluble or the breakaway type. The soluble support materials tend to dissolve when the object is placed in a suitable solvent solution. Whereas, the breakaway support materials require manual removal from the object. In other cases, certain polymers have a low tendency to stick to the surface of the platform, preventing them being used as matrices in 3D printing. To prevent such

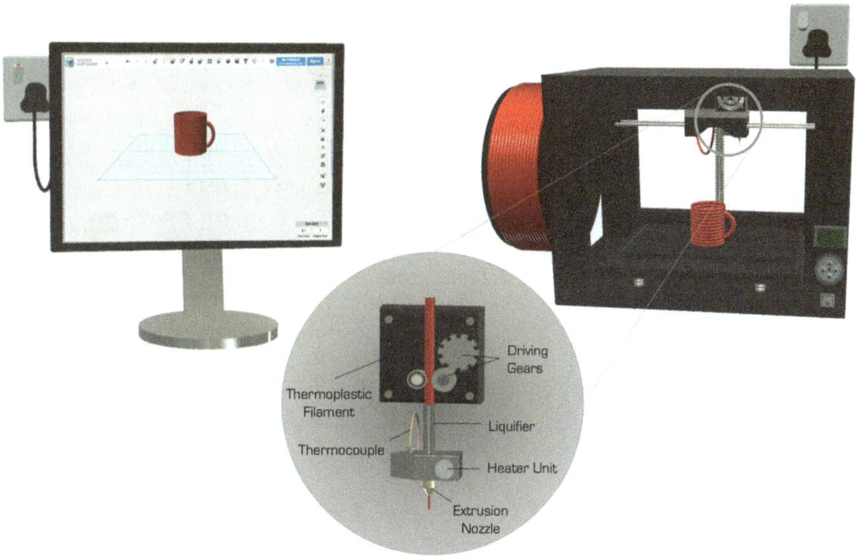

Fig. 6.1 3D Schematic representation of the FDM printing process

complications and to further enhance the adherence of the object, several solutions are suggested by experts. Suggestions include the heating of the platform of the printer, the use of an appropriate tape or piece of glass to cover the platform, or the use of a hairspray [5].

3 Industrial Applications

Figure 6.2 outlines a summary of the current advances in FDM 3D printing with emphasis on its applications in pharmacy. Due to its ability to incorporate a wide range of materials, FDM is considered one of the most diverse 3D printing technologies. Its uses extend to include the educational, mechanical, automotive, aerospace, robotics, maritime, entertainment, electronics, architectural, medical and pharmaceutical industries. Designers from all fields are able to imagine, sketch and see their final product come to life within minutes. Thus, the worlds of imagination and reality have been brought together within the same time frame.

This technology has the ability to create complex objects with unique characteristics such as high mechanical strength and thermal resistance. One key use of FDM is for the fabrication of prototypes, models and complete end products.

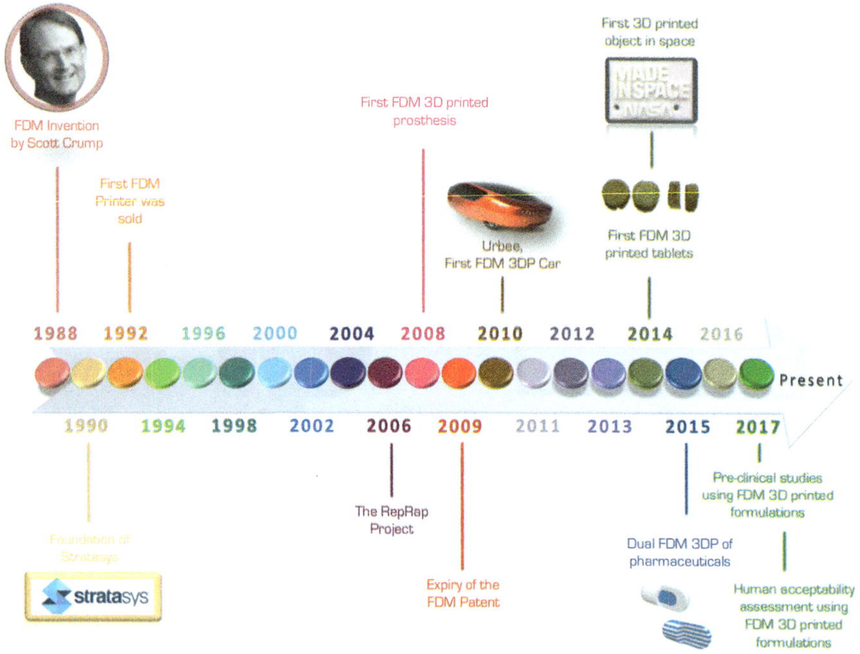

Fig. 6.2 Timeline from 1988 – present: Advances in the FDM technology in medicine and engineering. FDM: fused deposition modelling; HME: hot melt extrusion; 3DP: three-dimensional printing. (Images reprinted with permissions from [2, 6–9])

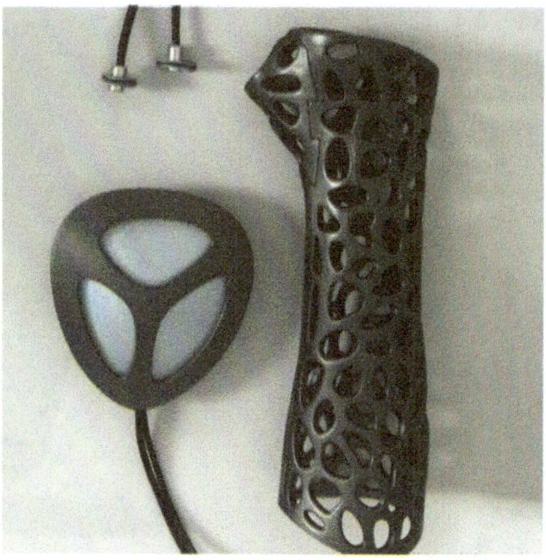

Fig. 6.3 Image of the osteoid cast. (Reprinted with permission from [16])

As such, FDM has been introduced as a contemporary tool for illustration in schools and universities, allowing students to visualise and interact with their tutors [10]. In medicine, surgeons have harnessed it to print anatomical guides and organs, specifically adapted to patients for training towards challenging surgeries [11]. The iLab organization in Haiti utilises this technology to manufacture on-demand medical devices such as umbilical cord clamps [12]. In dentistry, FDM is being used as a cost-effective method to create crowns, dentures, bridges and dental models [13]. Currently, this technology is being widely applied for the creation of customised prosthetic limbs to satisfy the needs of people from all ages, allowing them to live their lives normally [14]. Bespoke Inc. utilises FDM to prepare 3D printed fairings [15]; these stylish customised designs are used to cover prosthetics, lifting the self-esteem of patients.

Similarly, attachable osteoid casts have been also developed (Fig. 6.3) [16]. These functional devices utilise low-intensity pulsed ultrasound (LIPUS) to stimulate bone healing. This induces mechanical stress on the cells of fractured bones, reducing their healing duration by up to 38%. The casts are printed in two halves after which they are joined using a flexible pin. A 3D scanner, followed by modification using a CAD software, is utilised to adapt the cast to the patient.

In the automotive industry, several cars and motorcycles were 3D printed and expected to hit the market in the near future (Fig. 6.4) [6]. Several aircraft companies are even looking into using 3D printing to manufacture cost-effective, lightweight and heavy-duty airplane parts [17]. Currently the US military is exploring the ability to use the FDM technology to build robots, drones and even advanced weapons with unique characteristics otherwise unachievable to manufacture with other technologies [18].

Furthermore, not only has FDM proven to be a useful technique for building robust and durable architectural models and sculptures, it can even be used to build complete houses. In China, complete mansions were 3D printed using a recycled extrudate mixture in just a day. Subsequently, the same company 3D printed a complete office building in Dubai in just 17 days (Fig. 6.5) [19].

Fig. 6.4 Urbee, the first prototype car to have an entire 3D printed body. (Reprinted with permission from [6])

Fig. 6.5 The world's first 3D printed office in Dubai. (Reprinted with permission from [19])

In the near future, 3D printing is even expected to reach the moon or Mars! NASA has already started planning building complete human colonies using 3D printing. Their plan is to provide all necessities needed for human survival. The process they utilise is termed contour crafting, which is a hybrid division of FDM wherein concrete is extruded using automated robots. NASA has also partnered with MadeInSpace to create a 3D printer that operates in the absence of gravity called the Zero-G printer, enabling astronauts to print their own tools and equipment (Fig. 6.6) [20].

Nowadays, 3D printers can even be used to print other 3D printers. For instance, in 2005, a group of researchers at the University of Bath decided to create low-cost 3D printers initiating the RepRap project [21]. With the word RepRap being abbreviated from replicating rapid prototype, this project involved the use of an FDM

Fig. 6.6 The Zero-G 3D printer. (Reprinted with permission from [7])

printer that is able re-print all its parts to create other FDM printers. The project was successful and several versions of self-replicated RepRap printers were produced.

Overall, we can say that FDM is the future of marketing. We can imagine a world where people are able to design their own houses, furniture, decorations, accessories and even clothing using their personal FDM printers [22]. In other words, we will have our own instant shopping directory under our command.

4 3D Printing of Medicines

4.1 Hot Melt Extrusion

Although a wide variety of materials are compatible with FDM 3D printers, none of the commercialised filaments are regarded as safe for human ingestion. In pharmaceutical research, the feedstock of the FDM printers is created using a process called hot melt extrusion (HME). The first step of the extrusion process involves the preparation of the powder mixture. The required amounts of the polymer, drug and/or other excipients are carefully weighed and mixed. The acquired homogenous mixture is then poured into the extruder's hopper where it flows into the barrel. This powder is then melted at a controlled temperature which is slightly less than the melting temperature of the polymer. The molten material, with the aid of the rotating screw, high temperature and pressure, is then pushed out of a metal nozzle. The obtained filament is then deposited onto a collecting tray, where it is left to cool down. It is then sealed in appropriate packaging and stored in a vacuum desiccator to remove any trapped form of water or vapour.

A drug can be incorporated into filaments by two ways; either by being added into the powder mixture directly before starting the extrusion process or following the extrusion process, by submerging the filament into a suitable drug-containing solution, forcing the drug to passively diffuse into the filament. However,

Table 6.1 Summary of the current advancements on drug delivery systems fabricated using FDM 3D printing

Parameter	Components
Polymers	Polylactic acid (PLA), poly-ε-caprolactone (PCL), thermoplastic polyurethane (TPU), polyvinyl alcohol (PVA), polyvinyl pyrrolidone (PVP), methacrylic acid co-polymers (Eudragit – EUD; Grades: RL, RS, RL PO, E, EPO, L100, L100-55), polyvinyl caprolactam–polyvinyl acetate–polyethylene glycol (Soluplus), polyvinyl alcohol–polyethylene glycol (Kollicoat IR), galactoglucomanan, hydroxypropyl cellulose (HPC), hydroxypropyl methyl cellulose (HPMC), ethyl cellulose (EC), hydroxypropyl methyl cellulose acetate succinate (HPMC-AS; Grades LG, MG, HG), poly-L-lactic acid (PLLA), ethylene vinyl acetate (EVA).
Excipients (e.g. fillers, plasticisers and lubricants)	Triethyl citrate (TEC), magnesium stearate (MgSt), tricalcium phosphate, tribasic calcium phosphate (TCP), triacetin, talc, castor oil, oleic acid, lactose, tween 80, microcrystalline cellulose (MCC), polyethylene glycol (PEG), sodium starch glycolate, polysorbate, crospovidone, croscarmellose, polyethylene oxide (PEO), glycerol, methyl paraben, mannitol.
Dosage forms	Tablets, capsules, oral films, intra-uterine devices, subcutaneous devices, vaginal rings, topical wound dressings (e.g. nose and ear masks), beads, catheters, rods, microneedles, scaffolds, gastroretentive devices, implants.
Modes of drug release	Orodispersible, immediate release, sustained release, delayed release, pulsatile release.
Active pharmaceutical ingredients (APIs)	Paracetamol, theophylline, 4- or 5-aminosalicylic acid (4-ASA, 5-ASA), captopril, ibuprofen, prednisolone, budesonide, hydrochlorothiazide, caffeine, salicylic acid, dipyridamole, felodipine, furosemide, quinine, aripiprazole, indomethacin, nitrofurantoin, hydroxylapatite, rifampicin, isoniazid, glipizide, curcumin, progesterone, disulfiram, zinc, copper, silver, domperidone, gentamicin, methotrexate, deflazacort.

the diffusion method suffers from poor drug loading, wherein no more than 2% of the drug can be incorporated into the filaments [23, 24]. Therefore, combining HME with FDM was found to be more efficient.

Table 6.1 shows a summary of the current advancements in FDM 3D printing thus far. During the extrusion and printing processes, several complications might arise which make it difficult to obtain the desired formulation. For instance, some polymers suffer from poor flow properties, making their extrusion process slow and in some cases, rendering them unextrudable or unprintable. Therefore, the addition of a lubricant is essential for such polymers. It reduces the friction between the powder particles and the walls of the extruder allowing the mixture to flow in an easier and more consistent manner. As such, more control over the diameter of the extruded filament is obtained. Alternatively, the addition of a filler or the incorporation of other polymers could enhance the rheological properties of the mixture. Another problem often encountered during extrusion are brittle filaments wherein the brittleness prevents the filaments from being properly loaded into the printer's nozzle or causes them to break during the printing process. In some cases, this can lead to the blockage of the liquifier and result in the failure of the printing process. To prevent this, the addition of a suitable plasticiser is necessary. The plasticiser aids in reducing the glass transition temperature of the polymer and renders the filaments more flexible. However, choosing the right type and amount of plasticiser(s) is critical. This is mainly because excessive flexibility can cause the filament to bend inside the nozzle and block it.

For a filament to be printable it requires a diameter of 1.75 mm–3.00 mm [25]. However, some polymers tend to expand or shrink upon excessive heating, hence, the selection of a suitable nozzle size is critical. As such, nozzles with sizes ranging between 1.30–3.00 mm are usually used for extrusion. Alternatively, changing the distance between the tip of the nozzle and the collecting tray can also be utilised to adjust the diameter. A larger distance will produce thinner filaments, while a smaller distance will result in thicker filaments. This is mainly due to the pulling force associated with the effect of gravity.

The distinction of the extrusion process lies in its ability to incorporate different forms of excipients, including liquid and/or semi-solid materials. With low amounts of liquid materials, mixing using a mortar and pestle can be adequate to obtain a suitable consistency. However, where high amounts of liquid incorporation are needed, the liquid matrix has to be evaporated before starting the extrusion process [26]. This requires the mixing of all the excipients using an overhead stirrer. The mixture is then transferred into suitable containers such as Teflon petri dishes where it is then poured in thin layers. These are then kept in an oven and left to dry for a suitable period of time. Once the mixture dries, the obtained sheets are then removed and cut into small pieces which can then be used for extrusion. This can be seen as advantage over some of the other 3D printing processes where material restrictions may be a setback.

4.2 Pharmaceutical Applications

4.2.1 Oral Drug Delivery

Owing to its unique characteristics, extensive research has been conducted on the potential use of FDM in the pharmaceutical field. This technology has been utilised to prepare diverse 3D printed tablets (herein known as printlets: a term that we have coined to describe 3D printed oral dosage forms. Unlike conventional production methods, FDM is capable of producing printlets with various geometries and sizes, wherein changing the size and/or shape will result in a different drug release pattern [27]. As such, this demonstrates the potential ability of utilising this technology to print on-demand medications based on the patients' personal preferences. In the future, patients are expected to design their own tablets having their favourite shapes matched with their favourite colours.

Consequently, the first patient acceptability study conducted on 3D printed tablets was performed using FDM printlets. Results have shown that patients prefer swallowing different shapes opposed to others, thus affecting their compliance to

Fig. 6.7 (**a**) 3D designs and (**b**) images showing the different printlets geometries and sizes. (Reprinted with permission from [28])

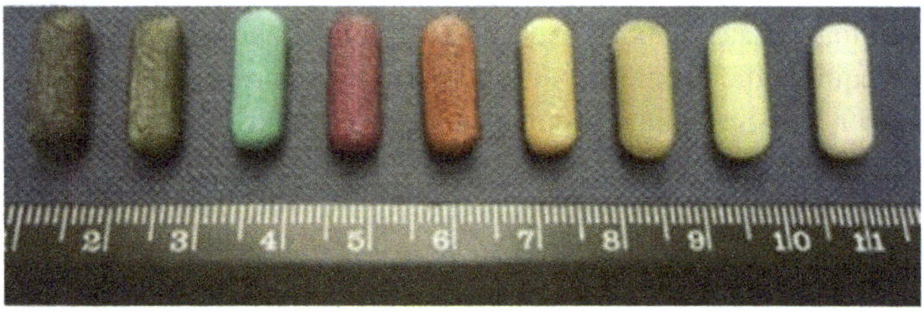

Fig. 6.8 Image showing printlets having different colours. (Reprinted with permission from [28])

medications (Fig. 6.7) [28]. This is perhaps driven by the patients' perception of ease of swallowing relative to the smoothness/sharpness of the edges of different printlets. As such, the tilted diamond shape scored the lowest and the torus shape the highest in terms of swallowing preference. In addition, it was evident that the size of the dosage form and its colour can also contribute to its acceptability (Fig. 6.8). This can be important when dealing with certain age groups such as paediatric and geriatric patients, wherein these factors can be significant determinants of the effectiveness of a treatment plan.

The flexibility of this technology allows the production of printlets exhibiting different drug release profiles, thus providing more control over the drug behaviour. For instance, Jamróz et al. utilised FDM to enhance the dissolution rate of aripiprazole through the fabrication of orodispersable films [29]. The physicochemical and mechanical characteristics of the 3D printed films were compared with that of their analogues produced using film casting. Although both films showed similar mechanical characteristics, results have shown that the 3D printed films were more flexible. However, the casted films required higher mechanical force to be broken, and thus, were considered more robust. In vitro dissolution studies have shown that

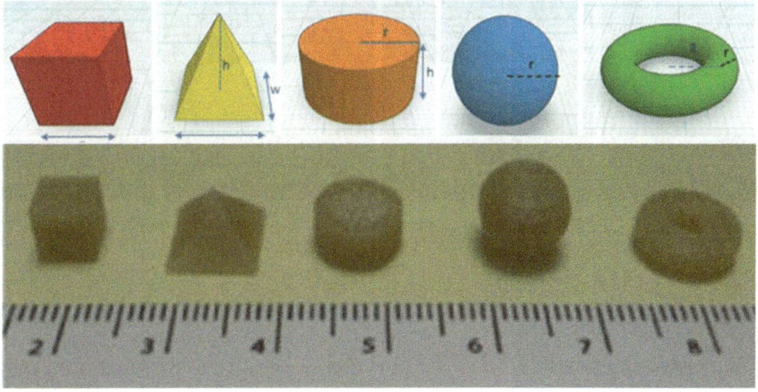

Fig. 6.9 (top) 3D designs and (bottom) images of printlets having different geometries, including (from left to right): cube, pyramid, cylinder, sphere and torus. (Reprinted with permission from [27])

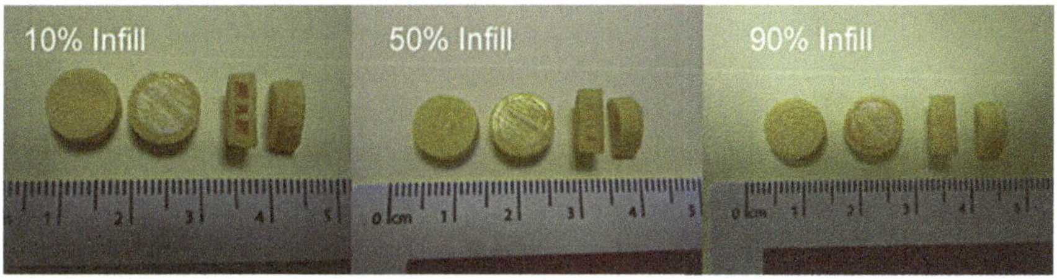

Fig. 6.10 Images of printlets having different percentage infill (from left to right): top, base, internal and lateral views. (Reprinted with permission from [21])

the 3D printed films exhibited an increased dissolution rate. This increase was thought to be due to the change in the physical state of the drug to the amorphous form during the extrusion and printing processes. Moreover, due to the printing pattern the FDM process follows, some pores were formed on the surface of the 3D printed films, which may have been another trigger for the accelerated action.

In a different approach, Goyanes et al. have shown that changing the geometry of the printlets can influence the drug release [27]. Five different shapes, including a cube, a pyramid, a cylinder, a sphere and a torus, were printed using FDM (Fig. 6.9). In vitro dissolution studies have shown that the ratio of the surface area relative to the volume of the printlets can alter the time needed for the drug to be completely released. Therefore, the fastest drug release was observed from the pyramid printlets, whereas, the slowest release was from the cylinder printlets.

In another study, it was concluded that the infill percentage is another determinant influencing the drug release (Fig. 6.10) [8]. When tablets having different infill percentages were printed, it was found that tablets containing less infill exhibited a faster drug release. Whereas, tablets with higher infill percentages showed extended release profiles. This was thought to be the result of exhibiting different swelling

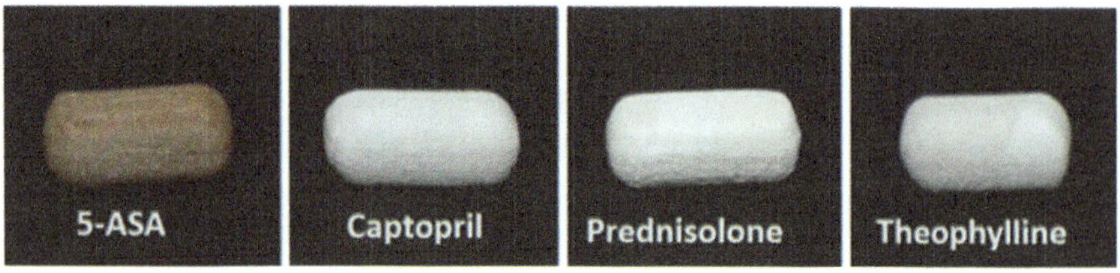

Fig. 6.11 Images of FDM caplets containing (from left to right): 5-ASA, captopril, prednisolone, and theophylline. ASA: aminosalicylic acid (Reprinted with permission from [32])

ratios, which was dependent upon the polymer concentration. On the contrary, in a study carried out by Chai et al., the change in infill percentage was exploited to create gastroretentive tablets [30]. Results have shown that the presence of enough amounts of gastric fluids can reverse the effect of changing the infill percentage that was seen in Goyanes et al. work. This is mainly due to the difference in densities, wherein, tablets having 0–20% infill had a density lower than that of the fluid media, causing them to float. The buoyancy effect increased the residence time of the tablets in the gastric region. On the other hand, tablets with higher infill percentages sunk and had a lower gastric residence time. The extension in residence time promotes the drug absorption from the early part of the small intestine, which is a primary site of drug absorption, enhancing the drug's bioavailability. This can be considered particularly significant for weakly basic drugs. However, such phenomenon is highly dependent upon a patient's diet and thus, a high variability in performance is expected.

To further demonstrate the versitality of this technology, modified release printlets were fabricated using different grades of enteric polymers [31]. They were successfully printed with a dose flexibility ranging from 5% up to 50% drug and having different infill percentages. In vitro dissolution studies revealed that all the printlets exhibited delayed drug release profiles, whereby less than 10% of the drug was released in the gastric region. Three factors, including polymer grade, drug dose and infill percentage, have contributed to this behaviour. In a different approach, Sadia et al. utilised FDM to fabricate immediate release caplets using Eudragit® EPO as the main polymeric matrix (Fig. 6.11) [32]. In vitro dissolution tests have shown that the complete drug release from all the caplets was achieved in the gastric region within less than 30mins. This was attributed to the ionisation of the polymer when in contact with the gastric media, giving rise to electrostatic repulsive forces. Thus, leading to the formation of gaps in the polymer chain, which in turn accelerated the drug release. Overall, it can be concluded that FDM printlets can be programmed to target pre-determined regions of the gastrointestinal tract, resulting in a more optimised treatment plan with higher drug efficiency.

Apart from targeting specific regions, FDM printlets can also be designed to release drug doses over defined periods of time. In 2017, Kempin et al. have demonstrated the importance of the polymer selection and drug loading on the drug activity [33]. Four different polymers, including Eudragit® RS (EUD RS), poly-ε-

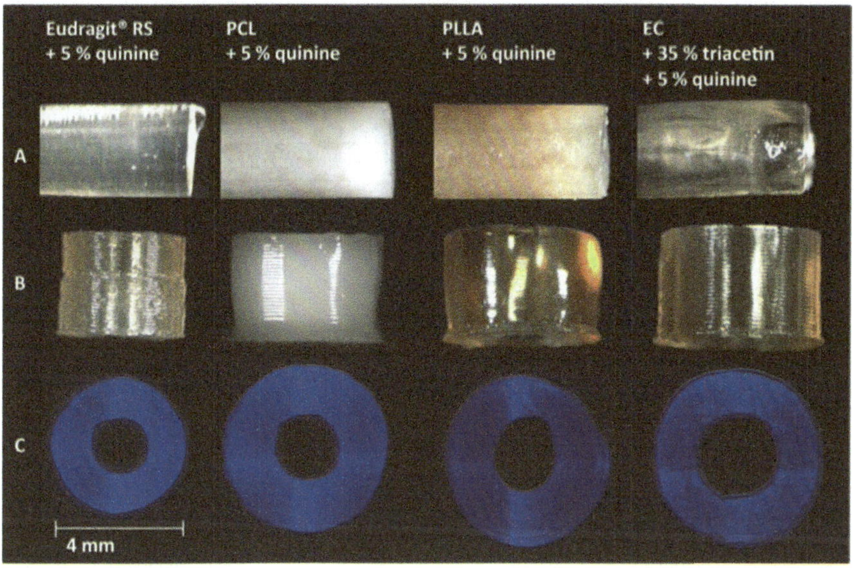

Fig. 6.12 (**a**) Reflected light microscopic images of the prepared filaments; (**b**) side view of the prepared implants; (**c**) top view of fluoresence microscopic images of the implants. (Reprinted with permission from [33])

caprolactone (PCL), poly(l-lactic acid) (PLLA) and ethyl cellulose (EC), were utilised in the study (Fig. 6.12). Quinine was incorporated into the individual polymers using solvent casting, whereby methylene chloride was utilised for PCL and PLLA and acetone was utilised for EUD RS and EC. In vitro dissolution studies have shown constant drug release over 58 days from the PCL implants, wherein ~76% of drug was released. In contrast, the Eudragit® RS and EC implants released no more than 3.7% and 4.5% over 78 and 100 days, respectively. As for the PLLA implants, up to ~53% of the drug was released over 177 days. Additionally, it was evident that a strong correlation lies between the drug loading and its release, wherein implants having higher loading have shown to release higher amounts of drug. Thus, it is clear that by selecting the appropriate polymer matrix and drug loading, one can tailor the duration of therapy based on the required dose regimen.

Although all the previously discussed factors have shown to alter the drug release, one factor, namely porosity, was found to have no effect. This was demonstrated through the work of Goyanes et al., whereby they prepared size 4 PVA caplets [34]. Four different formulations having 4.3% and 8.2% paracetamol and 4.7% and 9.5% caffeine were fabricated. SEM images have shown an increased porosity in the printed caplets compared to the parent filaments. This was attributed to the creation of gaps between the individual layers due to the printing process. Although once again in vitro dissolution tests have shown a strong correlation between the drug loading and its solubility, there was no apparent correlation between any of these

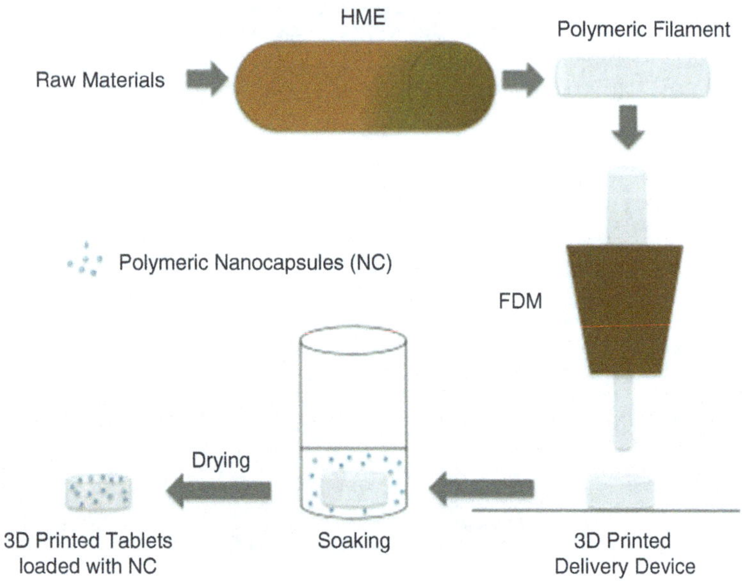

Fig. 6.13 Schematic representation of the process utilised to fabricate the nanocapsules-loaded printlets. (Reprinted with permission from [35])

two factors and the micropore volume. Thus, it can be concluded that the microstructure of the FDM printlets has no influence on the drug release.

The flexibility of the FDM technology is not only based on the versatility of the usable materials, but also on the fact that this technology can be combined with other technologies to fabricate unique products, otherwise unattainable with other processes. In 2017, Beck et al. combined FDM with nanotechnology to produce printlets incorporating drug-loaded nanocapsules [35]. They extruded four different filaments utilising PCL or Eudragit® RL 100 (EUD RL), whereby a channelling agent, namely mannitol, or an inert agent, namely microcrystalline cellulose, was incorporated into the mixture. The printlets were soaked in a nanocapsule suspension permitting the nanocapsules to diffuse into them (Fig. 6.13). Results have shown that several factors, including the polymer matrix composition, the infill percentage and presence or absence of a channelling agent, had a major impact on the drug loading and its release. The work successfully demonstrated the ability of FDM to transform the liquid nanocapsules suspension into a solid dosage form with enhanced drug loading.

The FDM printlets have high mechanical properties, making them more robust than compressed tablets. This high strength is still retained, even when hollow dosage forms are printed. This permits the incorporation of solid, semi-liquid or liquid materials with ease, without damaging the exterior structure. As such, Lim et al. utilised FDM to print hollow scaffolds that provide constant drug release [36]. The scaffolds consisted of a hollow base covered with a lid and were printed with multiple holes (Fig. 6.14). The holes were printed in different positions and in vary-

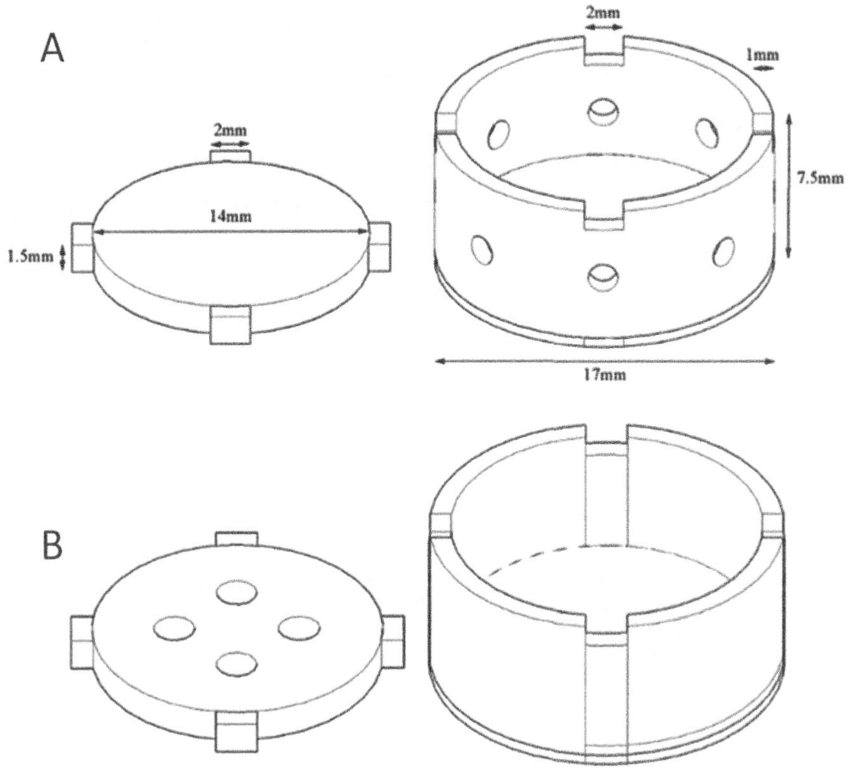

Fig. 6.14 3D design of the scaffold having (**a**) a lid and a body with holes on the side, (**b**) a lid with holes on top and a body. (Reprinted with permission from [36])

ing sizes to allow comparison. The scaffolds were filled with the drug powder and in vitro dissolution studies and high-performance liquid chromatography (HPLC) analysis were performed. Results have shown that the scaffolds having holes printed around the sides exhibited zero-order kinetics, indicating a sustained release profile. It was noted that the size of the hole acted as the rate-limiting step, whereby a faster drug release was observed in scaffolds with larger holes.

In 2015, Melocchi et al. successfully printed hollow capsule shells using hydroxypropyl cellulose (HPC) (Fig. 6.15) [25]. In vitro dissolution studies have shown that the 3D printed capsule shells exhibited characteristics similar to that of capsule shells produced using other techniques, such as injection moulding. As such, this allows the use of FDM for the on-site tailoring and filling of capsules as per patients' needs.

The work was followed by the separation of the capsular device into two distinct compartments [37]. The individual compartments were fabricated using different materials or having different thicknesses and were separated by middle joint part. Thus, the drug release from each compartment was initiated at different time points,

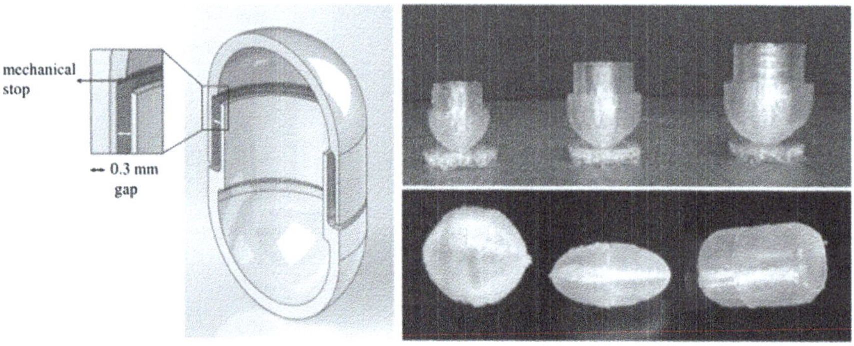

Fig. 6.15 Image showing (left) the 3D design and (right) 3D printed capsule shells compromising a cap and a body. (Reprinted with permission from [25])

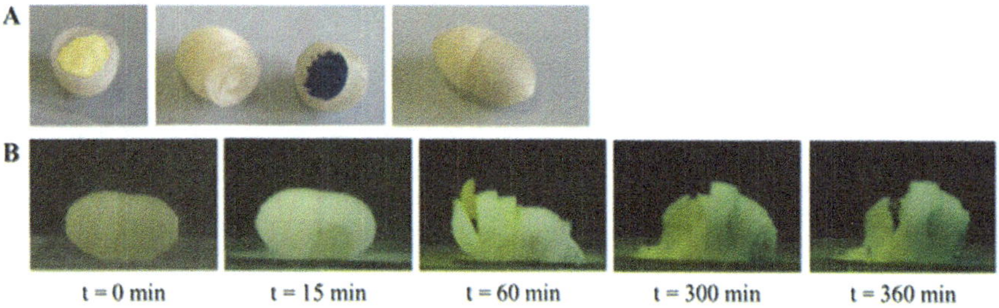

Fig. 6.16 Dual compartmental capsular shells having wall thicknesses of 600 and 1200 μm filled with yellow and blue dyes respectively (**a**) before and (**b**) during immersion in water. (Reprinted with permission from [37])

whereby the devices exhibited a two-pulse release. As shown in Fig. 6.16, the yellow dye was released after 60 min, whilst the blue dye's release was initiated after 4 h.

This invention is of high value for the creation of tailored printlets incorporating different drugs or drug doses, wherein it permits the use of drugs incompatible with one another. The dual compartments can be utilised to target different regions of the gastrointestinal tract whereby immediate release polymers can mediate to target the gastric region, whilst enteric polymers can be programmed to target intestinal regions. In late 2017, Genina et al. created a dual compartmental oral device using FDM for the treatment of tuberculosis [38]. The compartments were fabricated using PLA filaments, where one compartment was sealed with a water-soluble cap consisting of PVA whilst the other compartment remained unsealed (Fig. 6.17). A piece of filament containing either isoniazid (ISO) or rifampicin (RIF) was inserted into each compartment. Owing to the negative interaction upon the simultaneous use of these two drugs in low pH regions, this dual compartmental device was aimed at optimising the treatment plan by preventing the release of one drug in the gastric region. In vitro and in vivo studies have shown the ability of the devices to retard the drug release and its absorption when compared to the free filaments. Thus, it was

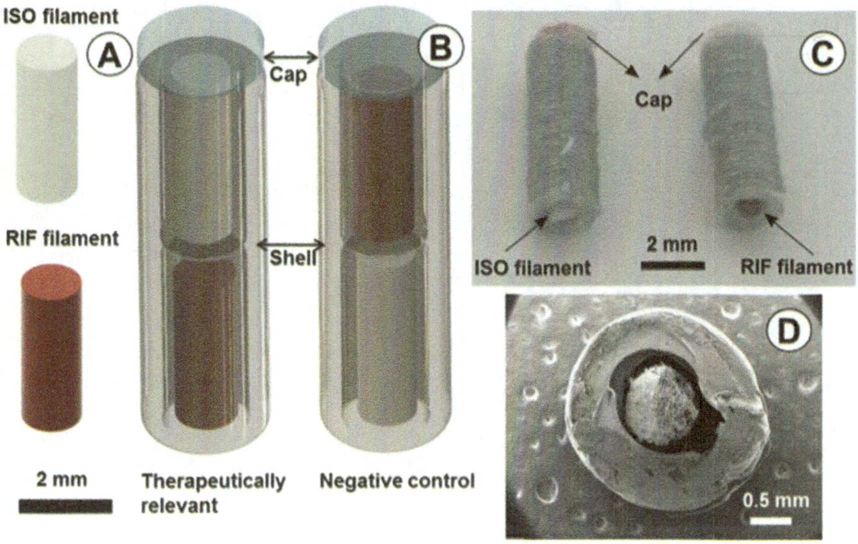

Fig. 6.17 (**a**) Schematic representation of the extruded filaments containing isoniazid (ISO) and rifampicin (RIF); (**b**) Schematic representation and (**c**) images of the 3D printed dual compartmental devices; (**d**) SEM image of an unsealed 3D printed compartment filled with a drug filament (top view). (Reprinted with permission from [38])

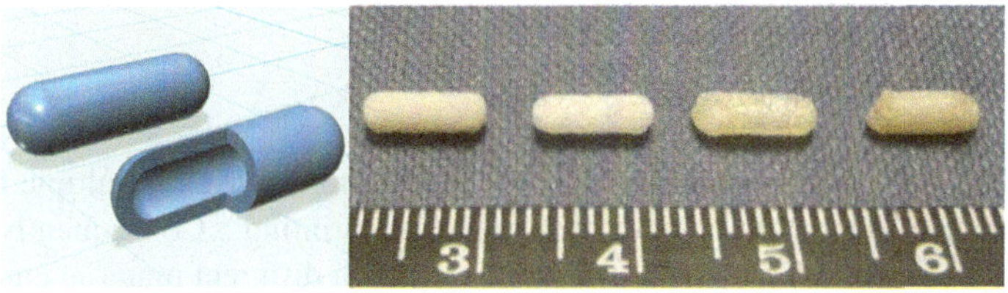

Fig. 6.18 Image showing (left) the 3D design and (right) printlets of (from left to right): Kollicoat IR, Klucel EF, Aqualon N7 and Aquasolve LG. (Reprinted with permission from [39])

evident that such approach is beneficial for reducing the interaction caused by this combination therapy.

In a different approach, FDM was utilised to fabricate printlets for in vivo testing in rats. Goyanes et al. printed size 9 hollow capsules using different types of polymers (Fig. 6.18) [39]. This study featured the first fused microPET/CT study performed in rodents using 3D printed devices. Due to the small size of the capsules, it was impossible to print the cap and body separately. Therefore, the capsules were printed as whole and then cut to allow the incorporation of the tracer. The two fragments were then re-attached using an adhesive agent. It was possible to print the capsules using different types of polymers, however, the printing resolution of the outer structure and internal holes varied. When tested, results have shown that none of the capsules were emptied from the stomach and each formulation exhibited a

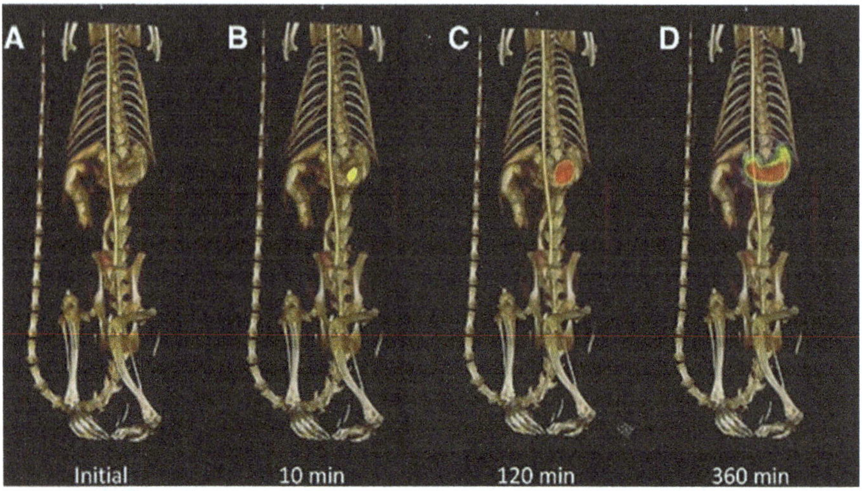

Fig. 6.19 Fused PET/CT image: (**a**) Prior to the administration of the Klucel EF printlets, (**b**) 10 min, (**c**) 120 min and (**d**) 360 min post-administration. (Reprinted with permission from [39])

different disintegration time, relative to the properties of the polymer (Fig. 6.19). This finding is quite important, showing that size 9 capsules are in fact only suitable for immediate release studies [40].

4.2.2 Dual Printing

Some FDM printers have multiple nozzles, permitting the use of multiple filaments simultaneously. This is known as "dual FDM 3D printing". Consequently, objects incorporating different materials can be printed. The different material can be used to print in distinct layers or can be combined within the same layer. Using dual FDM printing, the combination of several medications within the same dosage form can be attained, wherein the drugs can be dispersed in different types of polymer matrices or the printing pattern can be changed, generating different release profiles. This application can potentially increase patients' adherence to medications and lessen the burden of having to take several tablets within the same day. In 2015, Goyanes et al. used dual FDM printing to prepare multilayer caplets and duocaplets using paracetamol and caffeine as their model drugs (Fig. 6.20) [9]. In the multi-layer caplets, each layer consisted of one drug, with the following layer consisting of the other drug (Fig. 6.21a). The duocaplets, on the other hand, consisted of one drug being encapsulated within a shell made out of the other drug (Fig. 6.21b, c). In vitro dissolution studies have shown that drug release from the multilayer caplets was solely dependent on the characteristics of the individual polymers. Whereas in the duocaplets, drug release of the inner drug was driven by that of the outer drug. This concept was re-validated by Li et al. when they fabricated their duotablets, utilising glipizide as their active pharmaceutical agent [41].

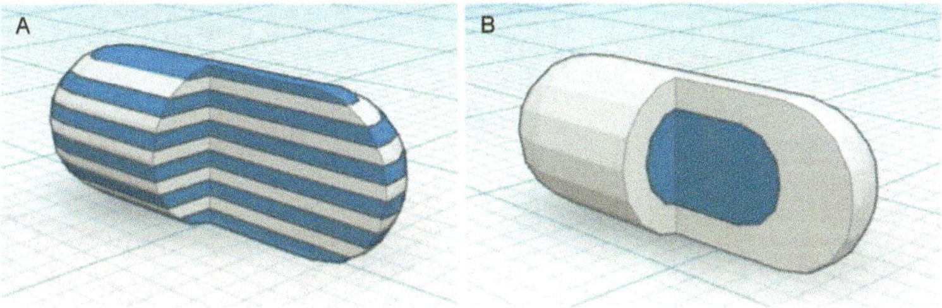

Fig. 6.20 3D designs of (**a**) the multilayered caplet and (**b**) duocaplet. (Reprinted with permission from [9])

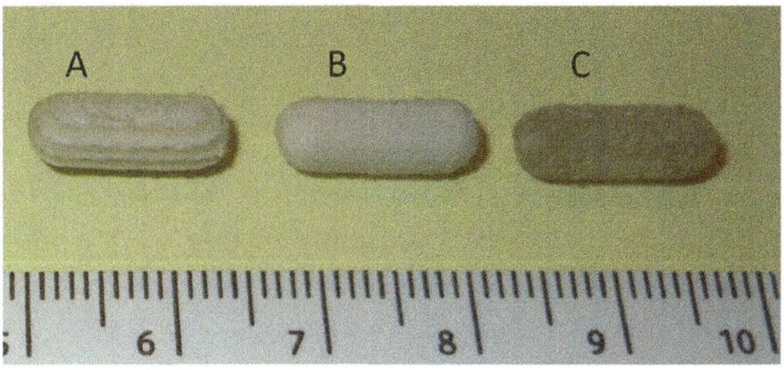

Fig. 6.21 Images of printlets: (**a**) multilayer paracetamol/caffeine caplet, (**b**) duocaplet of caffeine (outer) and paracetamol (core) and (**c**) duocaplet of paracetamol (outer) and caffeine (core). (Reprinted with permission from [9])

In another study, dual FDM printing was exploited for the fabrication of delayed release tablets [41]. In vitro dissolution studies revealed that the thickness of the outer layer and printing resolution are critical for the achievement of the required gastric resistance. As such, this can be utilised as an alternative method instead of coating.

In 2017, Gioumouxouzis et al. printed a triplecompartmental device having a middle hollow space [43]. The upper and bottom layers consisted of PLA where the inner core contained hydrochlorothiazide (HCTZ) dispersed within a PVA matrix (Fig. 6.22). These devices exhibited zero-order kinetics providing a constant drug action over the desired period of time.

With the constant development of FDM printers, printers nowadays have multiple nozzle heads that can operate together and replicate the same object simultaneously. Thus, they are expected to be useful for mass production of printlets in the future. Although these printers cannot produce the massive amounts that are normally produced using conventional tableting machines, they are still useful for the production of small on-demand batches of printlets [44]. Additionally, there are even printers that are able to combine several 3D printing technologies within the same printing job. For instance, the Hyrel printer allows the simultaneous use of the FDM and semi-solid extrusion technologies, thus broadening the pharmaceutical applications of 3D printing and permitting the development of more complex devices.

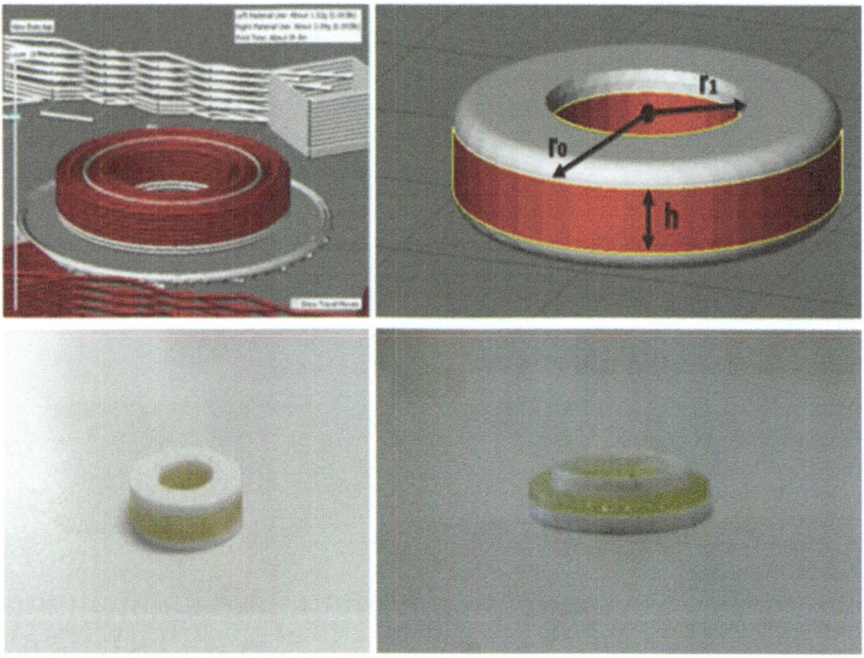

Fig. 6.22 Image showing (top) the 3D designs and (bottom) 3D printed three-compartment device exhibiting zero-order kinetics. (Reprinted with permission from [43])

4.2.3 Innovative Drug Delivery Devices

Apart from the oral dosage form printlets, FDM can also be used to produce bespoke drug delivery devices. In 2016, Goyanes et al. used FDM to prepare nose masks specifically tailored to the patient [26]. A 3D scanner was utilised to scan the patient's nose and the 3D design was generated. The aim of these masks was to provide a more optimised acne treatment using topical delivery. This was followed by a similar work conducted by Muwaffak et al., where they 3D printed anatomically tailored wound dressings in the shape of a nose and an ear (Fig. 6.23) [45]. The research group utilised metals including zinc, silver and copper as their antimicrobial agents to counteract the bacterial resistance seen with other anti-microbial agents. These masks enhanced the anti-microbial treatment without the need for systemic treatments. The adaptation of masks specific to patients helped in holding the dressings in the wound position, gaining further advantage over their analogous flat dressings.

In another study, Genina et al. utilised this technology to create personalised 3D printed intra-uterine and subcutaneous devices (Fig. 6.24) [46, 47]. In both studies, results have shown that the drug release was faster in the 3D printed devices compared to that from the extruded filaments. This was attributed to the presence of the

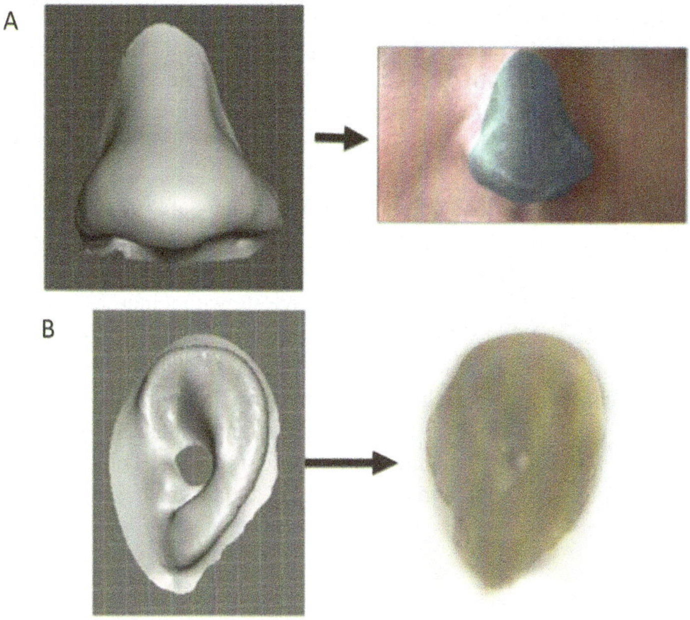

Fig. 6.23 (a) (left) 3D scan model of a nose (right) and its 3D printed wound dressing, (b) (left) 3D scan model of an ear (right) and its 3D printed wound dressing. (Reprinted with permission from [45])

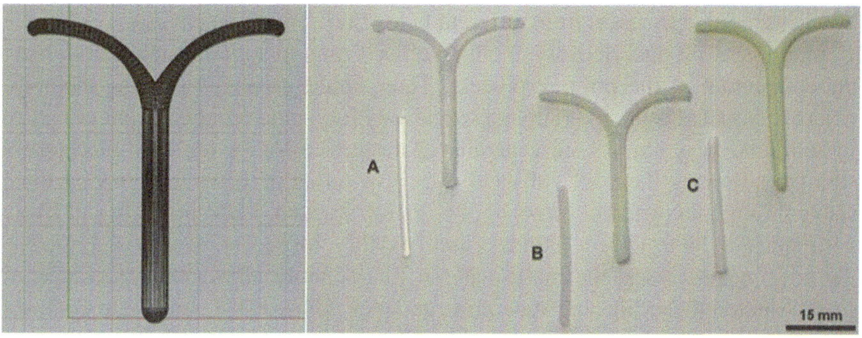

Fig. 6.24 (left) 3D design of an intra-uterine device; (right) the filaments and 3D printed intra-uterine devices containing: (a) no drug, (b) 5% indomethacin and (c) 15% indomethacin. (Reprinted with permission from [46])

drug in the amorphous form in the 3D printed structures, whereas the drug particles in the filaments were in the crystalline form.

In 2015, Weisman et al. demonstrated the feasibility of utilising FDM to fabricate 3D discs, beads and catheters [48]. Gentamicin and methotrexate were chosen as the model drugs, wherein they were combined with PLA to allow enhanced drug delivery and induce higher mechanical strength. When the drug-loaded filaments were compared with the 3D printed structures, results have shown that both exhibited similar anti-bacterial effects. Thus, this confirmed that the 3D printing process had no affect on the drug activity. The research group believed that their work demonstrated the ability of this technology to print on-demand anti-infective and chemotherapeutic drug delivery systems, providing more enhanced and optimised treatments.

Due to the strict regulations on patients' health and safety, only a few FDM 3D printed products are currently commercially available. This mainly includes anatomical surgical guides and artificial prosthesis [49]. However, following the approval of Spritam®, the regulatory bodies have shown an increased interest in 3D printing. Recently, the FDA has set up an Emerging Technology Team (ETT) consisting of a group of pharmaceutical experts to support and promote the regulatory evaluation of emerging manufacturing technologies, including 3D printing [50]. This was followed by the issuance of a guidance for the 'Technical Considerations for Additive Manufactured Devices" [51]. Thus, more FDM 3D printed products are anticipated to hit the market sooner than expected.

5 Advantages of FDM

Compared to other 3D printing technologies, FDM is considered to be one of the most feasible fabrication methods. This is mainly due to the low cost of the printers, user friendless of the technology and the high range of usable materials, making it highly accessible by users from different fields [52].

In terms of product delivery, the final objects tend to solidify rapidly following the completion of the printing process. Thus, unlike other 3D printing methods, objects fabricated using FDM do not require any form post-processing or curing [2]. In addition, these objects are generally characterised for having high mechanical strength, allowing them to withstand excessive forms of strain and making them stable throughout different forms of processing. As such, FDM printlets will likely not require to be tested for hardness and friability. Moreover, as opposed to some 3D printing processes, the multi-material functionality allows the integration of several materials within the same printing layer. Thus, FDM can create complex objects without the need for an assembly step, making it highly efficient [2].

6 Limitations of FDM

One of the major limitations of the FDM technology, is the high temperatures needed for extrusion and printing [54–58]. Therefore, extreme care should be taken when selecting the appropriate combinations of drugs and/or polymers. This is mainly because most polymers have high melting temperatures, at which most drugs start to degrade. Thus, some of the drug is lost during the extrusion and/or printing processes. This was clearly demonstrated in the 4-ASA printlets made by Goyanes et al., wherein more than half of the drug was found degraded following the printing process [23]. However this can be avoided through the careful selection of a drug-polymer combination, wherein the melting temperatures of the drug and polymer should be within the same range. In addition, the excipients added in the formulation can aid in optimising the formulation, thus reducing the extent of drug degradation. Currently, progress towards solving this issue has been made and several new research papers have reported printing at lowered temperatures (e.g. 40°C and 90°C) [59, 60]. With further research, maybe in the future we will be able to print at room temperature. Moreover, as the FDM-HME process is a duo-process, several factors will interfere with the overall printing process. Thus, unlike other 3D printing processes, printing with this technology is not straightforward. As such, perhaps there is a need for upgrading this procedure, such that the powder form of the formulation can be directly used for printing. This can reduce the time required for printing and lessen the extent to which the drug is subjected to heat.

7 Conclusion

The disease states of patients are specific to each individual. Therefore, generalising treatments can put some patients at risk, increasing the chances of treatment failure. Consequently, most pharmaceutical research is currently revolving around the personalisation of treatments specific to patients. This is the main urge behind researchers' interest in utilising 3D printing in the pharmaceutical field.

The uniqueness of the FDM technology lies in its ability to incorporate vast forms of materials. Moreover, the ability of FDM to integrate multiple materials within the same layer, allows it to be used to create the most complex structures that are impossible to obtain using conventional production methods. Hence, the applications of this technology in the pharmaceutical field are wide. The ease of use of this technology and the low space its feedstock occupies, makes it easy to be handled. In fact, compared to other 3D printing platforms, FDM is foreseen as having the highest potential to be used in fabricating on-site patient-specific dosage forms.

Nonetheless, based on the current research, the high temperatures associated with this technology can be regarded as its main drawback. This is a major issue when dealing with heat labile drugs, as it can potentially cause their degradation. Therefore, further research should be conducted in this area to reveal further polymers and/or excipients that can aid in alleviating this burden. In addition, since the technology consists of two steps, the hot melt extrusion and the printing process, the production process is considered to be rather time-consuming. Therefore, an appropriate solution might be creating printers specific for pharmaceutical use that allow the direct extrusion of the pharmaceutical blend during the printing process.

Overall, many aspects of this technology still remain vague and are yet to explored and optimised. Therefore, time will play a major role in determining its fate.

References

1. On3DPrinting.com. Inventor of 3D printing Scott Crump: "My Dreams Started in a Garage". 2013 [cited 2017 13-11]; Available from: http://on3dprinting.com/2013/09/17/inventor-of-3d-printing-scott-crump-my-dreams-started-in-a-garage/.
2. Stratasys. Make modern manufacturing happen. 2017 [cited 2017 03-12]; Available from: http://www.stratasys.com/.
3. Crump SS. Inventor; Google Patents, assignee. Apparatus and method for creating three-dimensional objects patent US 5121329 A. 1992 June 9, 1992.
4. Stratasys. FDM support removal. 2016 [cited 2017 12-10]; Available from: http://usglobalimages.stratasys.com/Main/Files/Best%20Practices_BP/BP_FDM_SupportRemoval.pdf?v=636008155408106896.
5. Horne R, Hausman KK. 3D printing for dummies. 2nd ed. Hoboken: Wiley; 2017.
6. Stratasys. Urbee the 3D printed car. 2013 [cited 2017 25-11]; Available from: http://usglobalimages.stratasys.com/Case%20Studies/Automotive/SSYS-CS-SSYS-Urbee%2007-13.pdf? v=635139938344953746#_ga=2.91568154.1616830113.1502104369-918546734.1477065242.
7. Made In Space. Leveraging space for industry. 2018 [cited 2018 04-01]; Available from: http://madeinspace.us/solutions-overview/.
8. Goyanes A, Buanz AB, Basit AW, Gaisford S. Fused-filament 3D printing (3DP) for fabrication of tablets. Int J Pharm. 2014;476(1–2):88–92.
9. Goyanes A, Wang J, Buanz A, Martinez-Pacheco R, Telford R, Gaisford S, et al. 3D printing of medicines: engineering novel oral devices with unique design and drug release characteristics. Mol Pharm. 2015;12(11):4077–84.
10. John M. How 3D printing is changing the shape of lessons. 2014 [cited 2017 28-10]; Available from: http://www.bbc.co.uk/news/business- 26871084.
11. Maxey K. 3D printed organs aid surgeons. 2013 [cited 2017 12-10]; Available from: http://www.engineering.com/3DPrinting/3DPrintingArticles/ArticleID/5575/3D-Printed-Organs-Aid-Surgeons.aspx.
12. Dara A. 3D printing Umbilical Cord Clamps in Haiti. Makezine 2013 [cited 2017 02-11]; Available from: https://makezine.com/2013/10/18/3dprinting- umbilical-cord-clamps-in-haiti/.
13. Jacob GS. Three-dimensional printing of dentures using fused deposition modeling. Inside Dental Technol. 2013;4(8):62–5.
14. E-NABLE. Enabling the future: a global network of passionate volunteers using 3D printing to give the world a "helping hand". 2017 [cited 2017 13-11]; Available from: http://enablingthefuture.org.
15. Summit S. Prosthetic limb with replaceable fairing. Google Patents 2010.
16. Osteoid. Osteoid. A better healing experience: Osteoid. 2017 [cited 2017 14-11]; Available from: http://www.osteoid.com/.
17. Wyman C. Stratasys additive manufacturing chosen by Airbus to produce 3D printed flight parts for its A350 XWB aircraft. Stratasys Blog 2015 [cited 2017 06-11]; Available from: http://blog.stratasys.com/2015/05/06/airbus-3d-printing/.
18. Peels J. 3D Printing in the Military. 2017 [cited 2017 23-09]; Available from: https://3dprint.com/165561/3d-printing-in-the-military/.
19. Dubai Future Foundation. The world's first 3D Printed Office. 2016 [cited 2017 28-10]; Available from: http://www.officeofthefuture.ae/.
20. Snyder M, Dunn J, Gonzalez E. The effects of microgravity on extrusion based additive manufacturing. AIAA SPACE 2013 Conference and Exposition. AIAA SPACE Forum: American Institute of Aeronautics and Astronautics; 2013.
21. Jones R, Haufe P, Sells E, Iravani P, Olliver V, Palmer C, et al. RepRap – the replicating rapid prototyper. Robotica. 2011;29(1):177–91.
22. Barnatt C. 3D printing. Third ed: Explaining The Future; 2016.

23. Goyanes A, Buanz AB, Hatton GB, Gaisford S, Basit AW. 3D printing of modified-release aminosalicylate (4-ASA and 5-ASA) tablets. Eur J Pharm Biopharm. 2015;89:157–62.
24. Skowyra J, Pietrzak K, Alhnan MA. Fabrication of extended-release patient-tailored prednisolone tablets via fused deposition modelling (FDM) 3D printing. Eur J Pharm Sci. 2015;68:11–7.
25. Melocchi A, Parietti F, Loreti G, Maroni A, Gazzaniga A, Zema L. 3D printing by fused deposition modeling (FDM) of a swellable/erodible capsular device for oral pulsatile release of drugs. J Drug Delivery Sci Technol. 2015;30(B):360–7.
26. Goyanes A, Det-Amornrat U, Wang J, Basit AW, Gaisford S. 3D scanning and 3D printing as innovative technologies for fabricating personalized topical drug delivery systems. J Control Release. 2016;234:41–8.
27. Goyanes A, Martinez PR, Buanz A, Basit A, Gaisford S. Effect of geometry on drug release from 3D printed tablets. Int J Pharm. 2015;494(2):657–63.
28. Goyanes A, Scarpa M, Kamlow M, Gaisford S, Basit AW, Orlu M. Patient acceptability of 3D printed medicines. Int J Pharm. 2017;530(1):71–8.
29. Jamróz W, Kurek M, Łyszczarz E, Szafraniec J, Knapik-Kowalczuk J, Syrek K, et al. 3D printed orodispersible films with aripiprazole. Int J Pharm. 2017;533:413–20.
30. Chai X, Chai H, Wang X, Yang J, Li J, Zhao Y, et al. Fused deposition Modeling (FDM) 3D printed tablets for Intragastric floating delivery of Domperidone. Sci Rep 2829. 2017;7(1)
31. Goyanes A, Fina F, Martorana A, Sedough D, Gaisford S, Basit AW. Development of modified release 3D printed tablets (printlets) with pharmaceutical excipients using additive manufacturing. Int J Pharm. 2017;527(1):21–30.
32. Sadia M, Sośnicka A, Arafat B, Isreb A, Ahmed W, Kelarakis A, et al. Adaptation of pharmaceutical excipients to FDM 3D printing for the fabrication of patient-tailored immediate release tablets. Int J Pharm. 2016;513(1):659–68.
33. Kempin W, Franz C, Koster L-C, Schneider F, Bogdahn M, Weitschies W, et al. Assessment of different polymers and drug loads for fused deposition modeling of drug loaded implants. Eur J Pharm Biopharm. 2017;115(Supplement C):84–93.
34. Goyanes A, Kobayashi M, Martínez-Pacheco R, Gaisford S, Basit AW. Fused-filament 3D printing of drug products: microstructure analysis and drug release characteristics of PVA-based caplets. Int J Pharm. 2016;514(1):290–5.
35. Beck RCR, Chaves PS, Goyanes A, Vukosavljevic B, Buanz A, Windbergs M, et al. 3D printed tablets loaded with polymeric nanocapsules: an innovative approach to produce customized drug delivery systems. Int J Pharm. 2017;528(1):268–79.
36. Lim SH, Chia SMY, Kang L, Yap KY-L. Three-dimensional printing of carbamazepine sustained-release scaffold. J Pharm Sci. 2016;105(7):2155–63.
37. Maroni A, Melocchi A, Parietti F, Foppoli A, Zema L, Gazzaniga A. 3D printed multi-compartment capsular devices for two-pulse oral drug delivery. J Control Release. 2017;268:10–8.
38. Genina N, Boetker JP, Colombo S, Harmankaya N, Rantanen J, Bohr A. Anti-tuberculosis drug combination for controlled oral delivery using 3D printed compartmental dosage forms: from drug product design to in vivo testing. J Control Release. 2017;268:40–8.
39. Goyanes A, Fernández-Ferreiro A, Majeed A, Gomez-Lado N, Awad A, Luaces-Rodríguez A, et al. PET/CT imaging of 3D printed devices in the gastrointestinal tract of rodents. Int J Pharm. 2018;536(1):158–64.
40. Hatton GB, Yadav V, Basit AW, Merchant HA. Animal farm: considerations in animal gastrointestinal physiology and relevance to drug delivery in humans. J Pharm Sci. 2015;104(9):2747–76.
41. Li Q, Wen H, Jia D, Guan X, Pan H, Yang Y, et al. Preparation and investigation of controlled-release glipizide novel oral device with three-dimensional printing. Int J Pharm. 2017;525(1):5–11.
42. Okwuosa TC, Pereira BC, Arafat B, Cieszynska M, Isreb A, Alhnan MA. Fabricating a Shell-Core delayed release tablet using dual FDM 3D printing for patient-centred therapy. Pharm Res. 2017;34(2):427–37.

43. Gioumouxouzis CI, Katsamenis OL, Bouropoulos N, Fatouros DG. 3D printed oral solid dosage forms containing hydrochlorothiazide for controlled drug delivery. J Drug Delivery Sci Technol. 2017;40:164–71.
44. Awad A, Trenfield SJ, Goyanes A, Gaisford S, Basit AW. Reshaping drug development using 3D printing. Drug Discov Today. 2018; https://doi.org/10.1016/j.drudis.2018.05.025.
45. Muwaffak Z, Goyanes A, Clark V, Basit AW, Hilton ST, Gaisford S. Patient-specific 3D scanned and 3D printed antimicrobial polycaprolactone wound dressings. Int J Pharm. 2017;527(1):161–70.
46. Genina N, Holländer J, Jukarainen H, Mäkilä E, Salonen J, Sandler N. Ethylene vinyl acetate (EVA) as a new drug carrier for 3D printed medical drug delivery devices. Eur J Pharm Sci. 2016;90:53–63.
47. Holländer J, Genina N, Jukarainen H, Khajeheian M, Rosling A, Mäkilä E, et al. Three-dimensional printed PCL-based implantable prototypes of medical devices for controlled drug delivery. J Pharm Sci. 2016;105(9):2665–76.
48. Weisman JA, Nicholson JC, Tappa K, Jammalamadaka U, Wilson CG, Mills D. Antibiotic and chemotherapeutic enhanced three-dimensional printer filaments and constructs for biomedical applications. Int J Nanomedicine. 2015;10:357–70.
49. U.S. Food and Drug Administration. Medical Applications of 3D Printing. 2016 [cited 2017 02–12]; Available from: https://www.fda.gov/MedicalDevices/ProductsandMedicalProcedures/3DPrintingofMedicalDevices/ucm500539.htm.
50. U.S. Food and Drug Administration. Emerging technology program. 2017 [cited 2017 12–11]; Available from: https://www.fda.gov/AboutFDA/CentersOffices/OfficeofMedicalProductsandTobacco/CDER/ucm523228.htm.
51. U.S. Food and Drug Administration. Technical considerations for additive manufactured devices. 2017 [cited 2018 12–01]; Available from: https://www.fda.gov/downloads/MedicalDevices/DeviceRegulationandGuidance/GuidanceDocuments/UCM499809.pdf.
52. Bikas H, Stavropoulos P, Chryssolouris G. Additive manufacturing methods and modelling approaches: a critical review. Int J Adv Manuf Technol. 2016;83(1):389–405.
53. Yan X, Gu P. A review of rapid prototyping technologies and systems. Comput Aided Des. 1996;28(4):307–18.
54. Trenfield SJ, Awad A, Goyanes A, Gaisford S, Basit AW. 3D printing pharmaceuticals: drug development to frontline care. Trends Pharmacol Sci. 2018;39(5):440–51.
55. Goyanes A, Chang H, Sedough D, Hatton GB, Wang J, Buanz A, Gaisford S, Basit AW. Fabrication of controlled-release budesonide tablets via desktop (FDM) 3D printing. Int J Pharm. 2015;496(2):414–20.
56. Awad A, Trenfield SJ, Gaisford S, Basit AW. 3D printed medicines: A new branch of digital healthcare. Int J Pharm. 2018;548(1):586–96.
57. Alhnan MA, Okwuosa TC, Sadia M, Wan KW, Ahmed W, Arafat B. Emergence of 3D printed dosage forms: opportunities and challenges. Pharm Res. 2016;33(8):1817–32.
58. Alomari M, Mohamed FH, Basit AW, Gaisford S. Personalised dosing: Printing a dose of one's own medicine. Int J Pharm. 2015;494(2):568–77.
59. Kollamaram G, Croker DM, Walker GM, Goyanes A, Basit AW, Gaisford S. Low temperature fused deposition modeling (FDM) 3D printing of thermolabile drugs. Int J Pharm. 2018;545(1–2):144–52.
60. Kempin W, Domsta V, Grathoff G, Brecht I, Semmling B, Tillmann S, et al. Immediate release 3D-printed tablets produced via fused deposition modeling of a thermo-sensitive drug. Pharm Res. 2018;35(6):124.

Chapter 7
The Role of Semi-Solid Extrusion Printing in Clinical Practice

Jack Firth, Abdul W. Basit, and Simon Gaisford

Abstract Residing under the umbrella term of 'material extrusion', semi-solid extrusion (SSE) 3D printing deposits gels or pastes in sequential layers to create a solid object. The physical nature of the feedstock allows SSE to print quickly at low temperatures with little compromise in accuracy. As a result, the technology has been extensively adopted in the field of bioprinting in which SSE can print living cells able to create large, complex structures of human tissue. In terms of pharmaceutical formulation, SSE is relatively unused. By building on the advancements made in other fields, the unique attributes of SSE printing are beginning to come to the fore. In particular, the ability of SSE to create complex tablets at a low heat means the technology has the potential to make on demand, personalised dosages within a clinical setting. As a result, the infancy of SSE in pharmaceutics should not be seen as a sign of inferiority but rather a real opportunity to bring 3D printing into the clinic.

Keywords Three-dimensional printing · Printing pharmaceuticals · Additive manufacturing · Personalized medications · Medical devices · Drug delivery systems

---

J. Firth (✉)
Department of Pharmaceutics, UCL School of Pharmacy, University College London, London, UK

Intract Pharma Ltd., London, UK
e-mail: jack.firth@intractpharma.com

A. W. Basit · S. Gaisford
Department of Pharmaceutics, UCL School of Pharmacy, University College London, London, UK

FabRx Ltd., Ashford, Kent, UK
e-mail: s.gaisford@ucl.ac.uk

A. W. Basit, S. Gaisford (eds.), *3D Printing of Pharmaceuticals*,
AAPS Advances in the Pharmaceutical Sciences Series 31,
https://doi.org/10.1007/978-3-319-90755-0_7

1 Introduction

Semi-solid extrusion printing (SSE) is a form of additive manufacturing based around the deposition of gels or pastes in sequential layers to create a solid object. Upon extrusion the semi-solids harden which allows the material to support the weight of subsequent tiers. The technique resides under the umbrella term of 'material extrusion' which began with the creation of fused deposition modelling (FDM). SSE however is differentiated by the use of print material that exists as a semi-solid at room temperature. Much like the FDM field, upon the expiration of Stratasys' patent over the technology, the growth of SSE development increased rapidly. Benefitting from the open-source community surrounding the 3D printing industry, SSE printers have been adapted for a multitude of uses from confectionery to living cells [1, 2]. The history of the technology is thus far from linear, evolving separately in each industry to create many variations in printer type and material with the only constant being the physical state of the feedstock used.

The physical nature of the feedstock allows SSE to print quickly at low temperatures with little compromise in accuracy. As a result, the technology has been advanced extensively in the field of bioprinting in which SSE can print living cells able to create large structures such as an aortic valve (Fig. 7.1) [3].

In terms of pharmaceutical formulation, SSE is a relatively unused technology with research only recently beginning to apply the process to tablet formulation. Nevertheless, by building on the advancements made in other fields, the unique attributes of SSE printing are beginning to be applied to drug development. Able to create complex tablets in a matter of minutes, it is thought that SSE opens up the possibility to make personalised dosages within a clinical setting, indicating its untapped potential for pharmaceutics [4, 5].

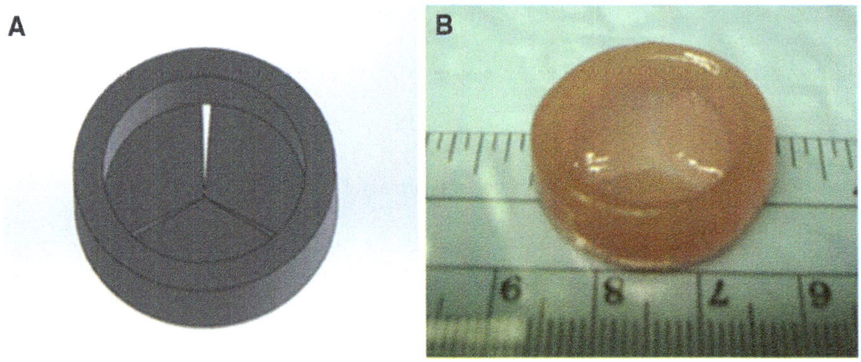

Fig. 7.1 (**a**) Heart valve original CAD design (**b**) Bioprinted aortic heart valve. Reprinted with permission from [3]

2 Pharmaceutical Semi-Solid Extrusion Printers

As the applications of semi-solid extrusion printing diversified into various industries, the urge to develop advanced printers became a necessity. With each industry having distinct requirements, the innovators implementing SSE into their respective fields have followed by adapting printers specifically for such uses. As a result, each field has developed their own form of SSE printer with some companies being dedicated to creating printers for a single industry. The Foodini™ printer (Natural Machines) for example has been used in high-end restaurants to create intricate, edible structures out of materials such as chocolate and guacamole [2]. In terms of pharmaceutical formulation, the technology is still in its infancy. As a result, researchers utilising the technique have had to adapt SSE devices from other fields to print pharmaceutical compounds. Two examples of such SSE printers that have been successfully adapted for use in pharmaceutical manufacturing are RegenHu's 3DDiscovery™ and the Fab@Home Model 2.

2.1 *3DDiscovery™*

Swiss-based RegenHU, primarily focuses on developing bioprinters, fluid dispensers and their proprietary matrix for 3D printing tissue (BioInk™ and OsteoInk™). The company has also developed its own bioprinting modelling software (BioTrack™ and BioCAD™) making it arguably one of the most integrated commercial biomanufacturing companies in the market. Priced at around $100,000 to $200,000, the 3DDiscovery consists of four extrusion print heads which use a pneumatic, syringe-based extrusion system (see Extrusion Sect. 3.1) (Fig. 7.2) [6]. Therefore, the machine is able to print multiple materials simultaneously to fabricate complex structures [7].

2.2 *Fab@Home Model 2*

Fab@Home was a project led by students at Cornell University's department of Mechanical and Aerospace Engineering, creating the first multi-material 3D printer available to the public and one of the first open-source DIY 3D printers. Since its release in 2006, the open-source nature of its design has allowed hundreds of Fab@Home 3D printers to be built across the world. Technological aspects of the printer have even been utilised in commercially-made printers, such as the Makerbot Replicator™. The project was closed in 2012, however this was preceded by the release of the Model 2, a cheaper and easier-to-use upgrade of the original printer.

The Model 2 is a mechanical syringe-based extrusion printer containing up to eight syringes which extrude via a single printer head (Fig. 7.3). The material kit for

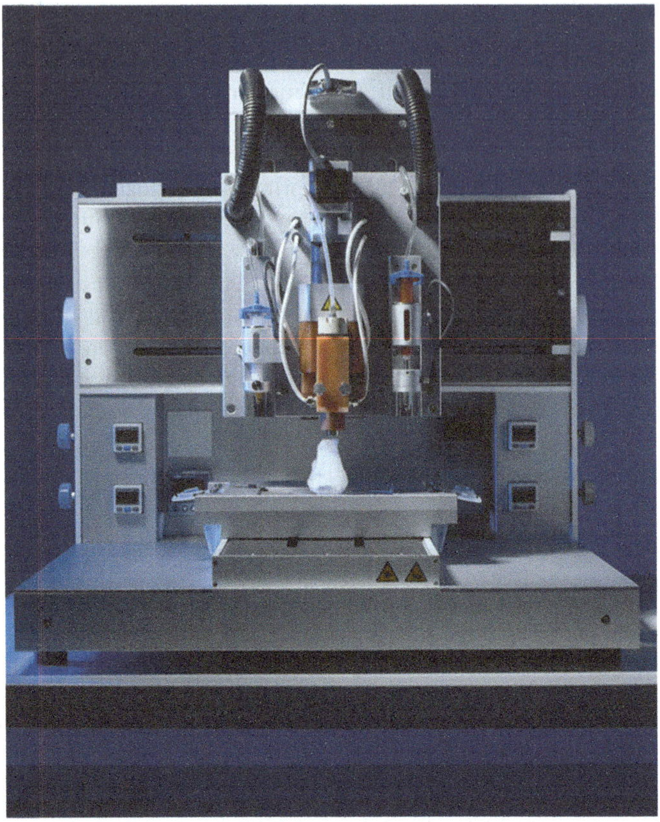

Fig. 7.2 RegenHU 3DDiscovery. Reprinted with permission from [6]

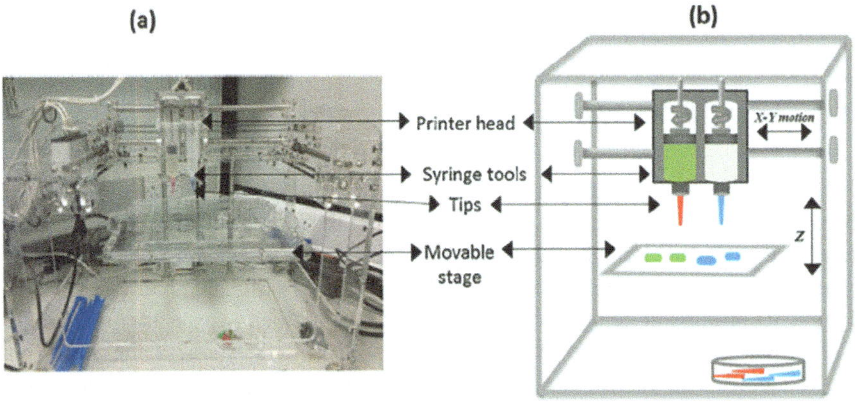

Fig. 7.3 (**a**) Photograph and (**b**) Schematic diagram of the Fab@Home printer model 2. For the purposes of the study, this machine was adapted to contain two printer heads. Reprinted with permission from [8]

the printer costs less than $1000 but does require assembly. In terms of printing, the Model 2 can reach print speeds of up to 20 mm/s and resolutions of 0.006 mm, 0.0035 mm and 0.0036 mm in the x-, y- and z-axis respectively [9]. The extrusion-based nature of the device means that a wide range of materials can be applied, and as a result of the open-source community surrounding the printer, materials such as silicone, biological materials, food and stainless steel have all been successfully printed using the Model 2.

3 Semi-Solid Extrusion Printing Process

At first glance the semi-solid extrusion printing process is similar to that of FDM; the material is heated, extruded and then solidifies to form the intended shape. Furthermore, as with any form of 3D printing, the desired structure is created using some form of computer aided design (CAD) programme which is then loaded into the printer software to enable object fabrication. Although it is clear to see the common underlying process between the two, there are a number of contrasting mechanisms. The differences primarily revolve around the feedstock used, particularly the semi-solid nature of the material.

3.1 Extrusion

The defining aspect of an SSE printer is the extrusion process. Unlike other types of 3D printing, the material begins as a gel or paste and is printed using a form of syringe. As the material is contained in a syringe, the heating process can occur both internally, immediately prior to printing or externally, perhaps whilst another syringe is being used. However, care must be taken when heating as it can cause the material to become too liquefied to hold any structure upon deposition. Optimisation is thus required to identify the ideal temperature needed to print using different materials. The extrusion of the material itself can be driven by either a (1) pneumatic, (2) mechanical, or (3) solenoid-based system [10].

1. **A pneumatic-based system** utilises air pressure to depress the syringe which then extrudes the product (Figure 7.4a). This mechanism can often use a valve on the nozzle to ensure material is not released when no pressure is applied. However, the hydrogels used in pharmaceutical SSE printing can rely on a valve-free model due to their increased viscosity. These systems have been associated with a delay between initiating, dispensing and extrusion, a consequence of using compressed gas. It can therefore be difficult to ensure the accuracy of material dispensed when using small volumes. The method is seen as optimal when using highly viscous materials as the gas is able to reach significant pressures without compromising the mechanical integrity of the extrusion system.

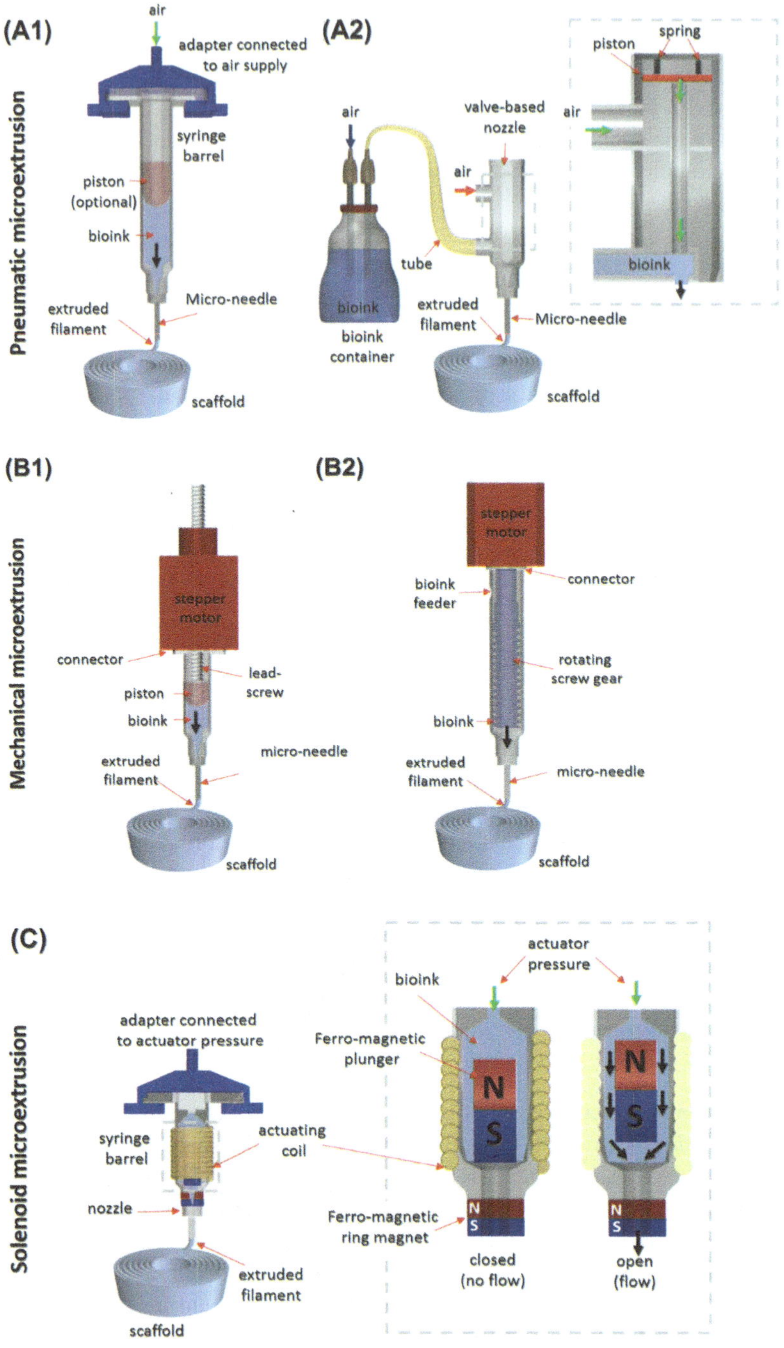

Fig. 7.4 Extrusion-based bioprinting systems: (**a**) pneumatic microextrusion including (a1) valve-free and (a2) valve based, (**b**) mechanical microextrusion including (b1) piston- or (b2) screw-driven, and (**c**) solenoid microextrusion. Reprinted with permission from [10]

2. **Mechanical-based systems**, as the name suggests, applies mechanical force directly to the syringe (Figure 7.4b). These systems can take either a piston or screw-based configuration to control gel extrusion, overall providing a much simpler mechanism than the pneumatic process. The portability of these systems are also much improved, lacking the requirement for a cumbersome compressed gas apparatus. Piston-based printing is seen to provide more control over the extrusion flow, as the screw-based approach can suffer from pressure changes. Conversely, the screw-based mechanism offers an increased level of spatial control and is able to print semi-solids of a higher viscosity than with a piston.
3. **Solenoid micro-extrusion printers** apply electrical pulses to a valve located at the base of the syringe (Figure 7.4c). In doing so, the pulse terminates the magnetic attraction generated between a floating ferromagnetic plunger and a ferromagnetic ring. This mechanism allows the printers to dispense sub-μL volumes of material. Therefore, solenoid printers are not currently used in pharmaceuticals due to the viscosity of the gels/pastes applied. The mechanism is sometimes limited in terms of its reproducibility and is primarily used in bioprinting. In some printers a time-lapse between actuation (where the coil is energised) and the valve opening can affect the consistency of the volumes extruded, resulting in an uneven deposition of material.

3.2 Printing

Upon initiation of the printing process, the semi-solid material is extruded through a nozzle at the base of the syringe. It is here where the viscosity of the material can significantly impact the process; if the material is too liquefied, the structure cannot hold its shape. If it is too viscous, it can clog the nozzle and potentially damage the extrusion mechanism [10]. To rectify this, the formulation of the material can be adjusted to alter its physical properties and increase the liquidity. Alternatively, it is possible to reduce the viscosity of the feedstock by heating the syringe, or by using a larger nozzle head able to avoid clogging. It should be noted that in the case of a formulation not being viscous enough, these same processes could also be applied in reverse. It will come as no surprise that the positioning of the extrusion head is of vital importance when printing. However, it is also necessary to calibrate the distance from the bedplate to the viscosity of the material printed (in some cases this is completed automatically). If the distance from the bedplate is too great, a more viscous solution will accumulate and solidify at the tip of the nozzle; if it is too close, the continuous extrusion will be broken by the nozzle against the bedplate, leading to a streaky, incomplete deposition [10]. Once the head is calibrated, the printer, often pre-programmed to do so, will print an arbitrary shape in order to ensure a continuous flow of material, which is often initially interrupted by air bubbles accumulating in the nozzle of the syringe.

Positioning of the nozzle is orientated via a computer-controlled motor system, allowing the printer head to move via the x, y and z axis. In some cases, the printer bed can move vertically, rather than the nozzle; this is however not the case in the printers thus far adopted for a pharmaceutical purpose. The bed can also cause a number of issues when printing. In order for the layers of material to be deposited, the initial extruded layer must not only adhere to the bed, but also must be able to be easily removed once the process is complete. To facilitate this, the print bed can be heated to limit the stress in the base layers, thus minimising warping of the material, which can lead to premature release of the printed part. However, once again this temperature needs to be carefully optimised to ensure the material is able to solidify sufficiently and thus retain its shape.

4 Pharmaceutical Applications of Semi-Solid Extrusion

In terms of pharmaceutical formulation, SSE is a relatively novel concept, the first application only occurring in 2015 [8]. As a result, the Roberts group at the University of Nottingham and the Pan group at Shenyang Pharmaceutical University are in fact the only teams to have published research using this technology in tablet development, though other groups are known to be actively researching the area [11, 12].

4.1 Controlled Release Bilayer Tablets

To assess the validity of SSE in tablet fabrication, the Roberts group sought to create controlled release bilayer tablets containing guaifenesin and subsequently compare the tablets to the commercially available product (Fig. 7.5) [8]. Used clinically as an expectorant, guaifenesin is available as a bilayer tablet consisting of immediate and sustained release layers. The bilayer tablets provide a rapid administration via burst release while simultaneously maintaining therapeutically effective levels of drug release over an extended period. To achieve these distinct layers, two HPMC gels were prepared composed of different viscosity grades of HPMC; HPMC 2910 for the immediate release layer and HPMC 2208 for the sustained release layer. Printing of the tablets was then completed by the Fab@Home Model 2 printer using an 800 μm nozzle head.

The dissolution profiles of the 3D printed tablets successfully displayed both immediate and sustained releases of guaifenesin as intended. It was seen that the formulations containing the highest concentration of HMPC (14%) most closely reflected the drug release profiles of the commercial tablets (although no higher concentrations were investigated). The concentration of HMPC, and in turn the viscosity of the gel base, was therefore integral in determining the drug release

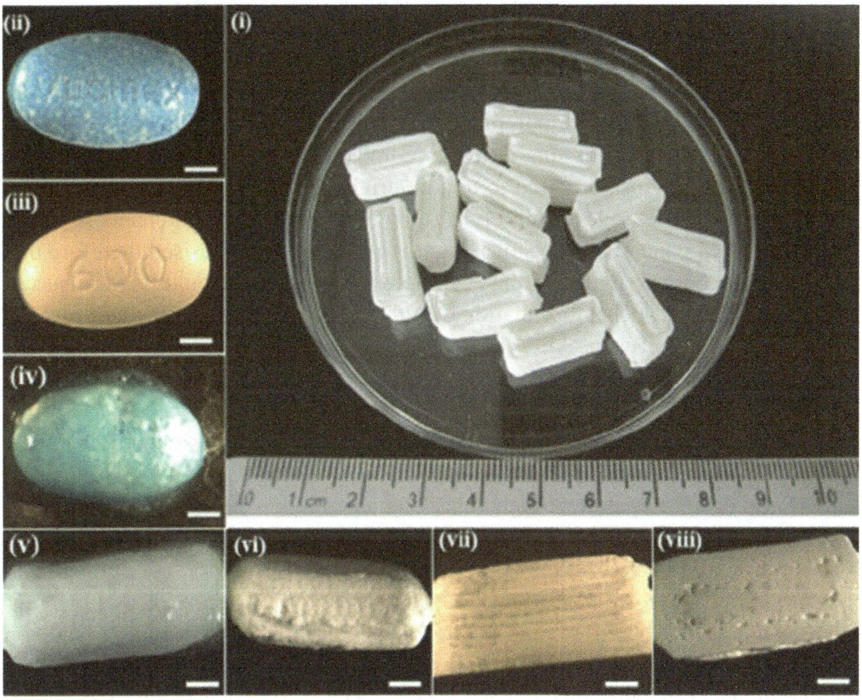

Fig. 7.5 (i) 3D printer produced guaifenesin bilayer tablets GBT-HPMC (14%, w/w). (ii) Top view and (iii) underside view of commercial GBT. (iv) and (v) Gel barrier surrounding commercial GBT and 3D printed bilayer tablet after 2 h dissolution test, respectively. (vi) Top view, (vii) side view and (viii) underside view of individual 3D printed bilayer tablets. The scale bar (ii–viii) is 10 mm. Reprinted with permission from [8]

profiles. This therefore implies the potential for adjusting the formulation based upon the specific profile required.

The researchers did note that tablets with a lower HPMC concentration showed significantly higher in-batch size variation, suggesting that a less viscous gel struggles to ensure uniformity in printed shape. The rigidity of the tablets differed greatly, with the standard formulations displaying a hardness value over twice that of the 3D printed tablets. Compression is not part of 3D printing process, and therefore it was expected that the 3D printed tablets would not have an equivalent hardness. This is particularly the case using SSE, as such formulations are based on gel material and are thus unlikely to harden to the same level when compared with compressed powders. It should be noted that it is possible to control the tablet hardness by other means than compression. This may include increasing the concentration of the binder HPMC or adding dry binder (e.g., PVP). However, as all the printed tablets could be handled readily without any loss of structural integrity, it seems unnecessary to seek a harder tablet, particularly as it could affect drug release.

4.2 Three-In-one Polypill with Defined Release Profiles

Building upon the successful creation of a bilayer tablet, SSE was subsequently shown to fabricate a complex polypill (a tablet containing a combination of several compounds) incorporating the drugs nifedipine, captopril and glipizide [13]. Nifedipine and captopril are used to treat arterial hypertension in type II diabetics, whilst glipizide encourages insulin release by blocking potassium channels within the beta cells of pancreatic islets of Langerhans. It was hypothesised therefore that a single tablet containing nifedipine, captopril and glipizide would be of great use to diabetic patients; such a formulation would however require specific independent release profiles for each API. This level of complexity is challenging with conventional tablet manufacturing processes, but with the ability to extrude multiple feedstocks at once, SSE provides an appropriate solution.

The 3D printed tablets created by the Roberts group were designed to involve a multi-compartment strategy ensuring the APIs were separated and could therefore achieve the desired control of their release kinetics independently (Fig. 7.6). SSE was used to fabricate a base compartment containing captopril, fully encapsulated by a porous cellulose acetate shell, so as to restrict drug release through osmosis and thus providing a more controlled release profile. The top layer consisted of two segments containing either nifedipine or glipizide, encapsulated by the porous shell on all but the superior face to enable drug release to be governed by diffusion. Both compartments were also mixed with HMPC. This resulted in a more sustained release mechanism as HMPC swells and forms a viscous gel-like layer on contact

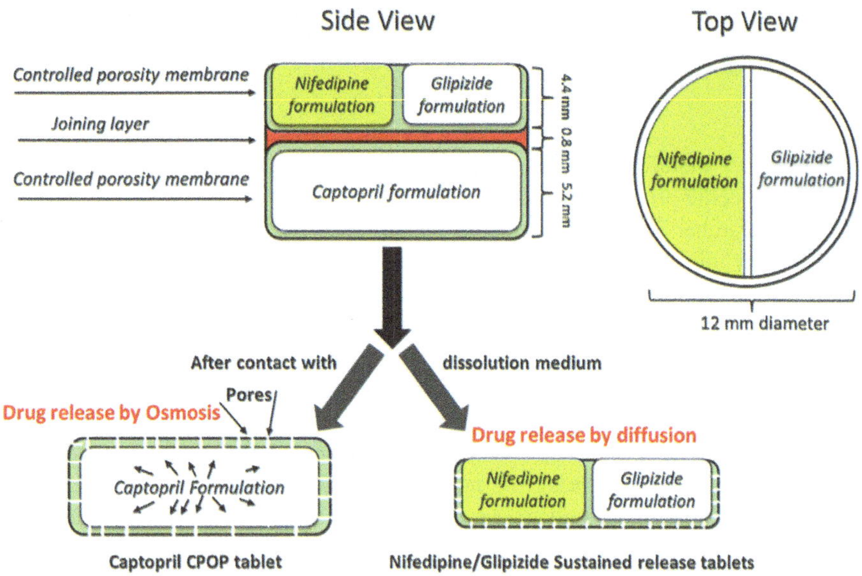

Fig. 7.6 Schematic diagram of the Roberts group multi-compartment tablet design. Reprinted with permission from [13]

with aqueous liquid. Consequently, the gel layer creates a diffusion barrier, inhibiting water from reaching the drug and thus reducing drug release by diffusion or erosion. The two compartment levels were connected via a 'joining layer' designed to disintegrate quickly within an hour, allowing the tablet to split into two parts with different drug release mechanisms [13].

Extrusion was performed using a RegenHu 3DDiscovery™ printer, able to extrude the five separate feedstocks simultaneously, therefore allowing the tablet to be created in a single print. As the viscosity of the feedstock is critical when using SSE, the optimisation of the cellulose acetate shell paste was crucial. This was achieved by adjusting the ratios of cellulose acetate and D-Mannitol together with the plasticiser, PEG 6000 [13]. Increasing the cellulose acetate concentration increased the paste viscosity whereas increasing the content of D-mannitol in the shell produced a very fragile solid. Upon printing, the 3D printed tablets were placed in a vacuum dryer at 40°C for 24 hours to dry completely. These post-printing drying processes are required when using certain gels to ensure that the 3D printed tablets have sufficiently solidified.

4.3 Multi-Release Polypill

Polycap™, manufactured by Cadila Pharmaceuticals Ltd., is a therapy for cardiovascular disease and the only polypill formulation commercially available [14]. The Polycap™ capsule contains pravastatin, atenolol, ramipril, aspirin and hydrochlorothiazide in five separate compartments each with distinct release profiles. Current manufacturing techniques require the Polycap™ to be created with predetermined doses of each API. SSE however allows the polypill to be tailored both in terms of dosage and release profiles simply by altering the geometric shape and size of each component, as well as the type of barriers surrounding them. Before adopting SSE, similar studies were previously completed using powder bed printing; however these tablets were found to require extensive drying post-printing as well as being highly friable [15].

The Roberts group applied SSE printing using a RegenHu 3DDiscovery™ (Fig. 7.2) to first create a hydrophobic cellulose acetate shell. Within this, the active drugs atenolol, pravastatin, and ramipril were each combined with a hydrophilic matrix (HPMC) and were extruded into the segments to form sustained release compartments (Fig. 7.7) [16]. To create immediate release compartments containing aspirin and hydrochlorothiazide, the compounds were mixed with sodium starch glycolate (a disintegrant) and extruded on top of the sustained release segments. Printing was performed using a 500 μm print tip and took 25 minutes followed by drying in a vacuum dryer at 40°C for 24 hours.

The group successfully demonstrated that SSE printing is able to create a complex polypill tablet with multiple release profiles. Analysis confirmed that the tablet was stable and that printing did not compromise the integrity of the APIs whilst the dissolution profiles acted as intended. By utilising SSE, the group were

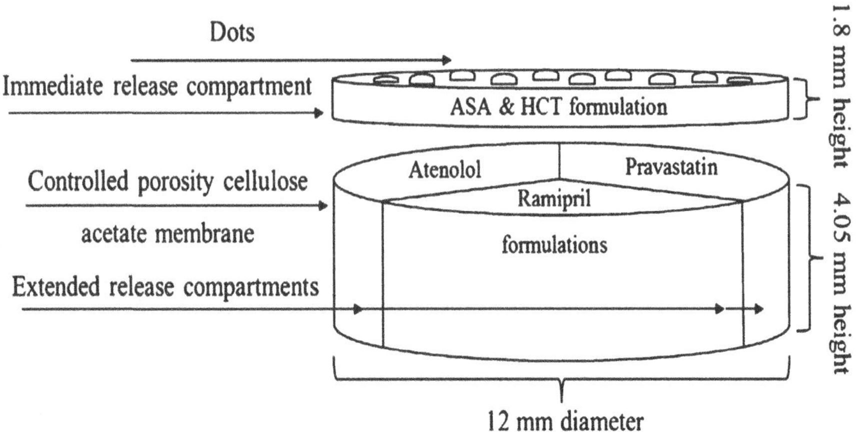

Fig. 7.7 Schematic diagram of the Roberts group polypill design. Reprinted with permission from [16]

able to validate the technology for complex, intricate formulations. It is also of note that the group elected to re-use the RegenHu 3DDiscovery™, suggesting that the Fab@Home Model 2 is limited in its applications, whilst the RegenHu printer is preferable when creating more complex 3D printed tablets.

4.4 Gastro-Floating Tablets

Rather than seeking to develop multi-release polypills, the Pan group adopted SSE in order to fabricate a novel form of gastro-floating device [17]. Used as a gastroretentive drug delivery system, such formulations are intended for drugs that locally take effect in the proximal part of gastrointestinal tract, are unstable in intestinal fluid, or are poorly soluble in the intestinal tract. These gastro-floating tablets are designed to float on the surface of the gastric contents, avoiding passage into the duodenum until the formulation is completely dissolved. SSE was therefore adopted to fabricate 3D printed tablets of dipyridamole with a low-density lattice internal structure in order to provide buoyancy. As a result, the formulation was theorised to improve the dipyridamole release rate and consequently improve bioavailability and therapeutic efficacy of the drug.

Using a MAMII extrusion printer (Fochif Mechatronics Technology Co., Ltd.) the structures were able to be created at room temperature however they did require a 12 hour drying process at 40°C. In order to create gastro-floating formulations, the group created 3D printed tablets with various infill percentages (30%, 50%, 70% and 100%). By manipulating the design of the lattice structure, it was possible to vary the degree at which the core of the structure was filled with material (Fig. 7.8). Consequently, the formulations with a lower infill percentage possessed a reduced density and a larger amount of encapsulated air than a conventional filled tablet.

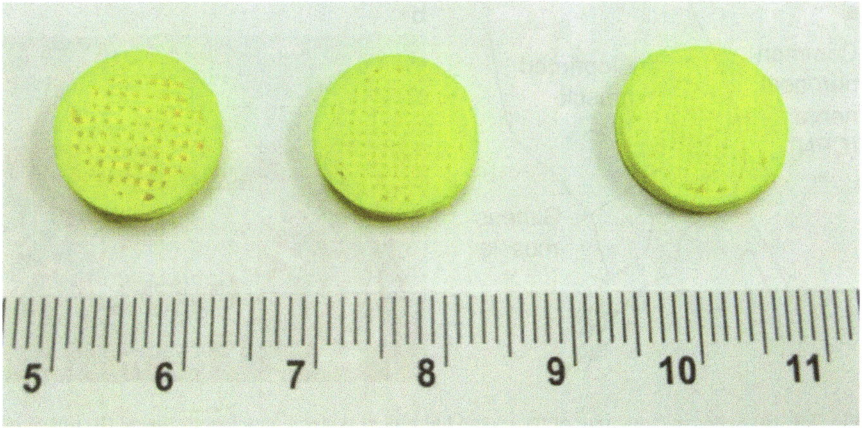

Fig. 7.8 Gastro-floating 3D printed tablets with 30%, 50% and 70% infill from left to right. Reprinted with permission from [17]

As a result, the low density 3D printed tablets were able to remain afloat in simulated gastric fluid with the buoyancy time of the 3D printed tablets negatively correlated with infill percentage.

By successfully creating a novel form of gastro-floating device, the Pan group clearly demonstrated the versatility of SSE. In being able to make intricate adjustments to the internal lattice structure of the formulations, it is not hard to envisage such gastro-floating devices being tailored to each patient dependent upon their variations in gastric physiology.

5 Other Applications of Semi-Solid Extrusion

Though SSE printing is in its infancy within the pharmaceutical sector [11], the technology has been widely adopted in other areas of healthcare. In the field of bioengineering for example, SSE has been honed and optimised for use, becoming an industry standard. This demonstrates that although its applications in the pharmaceutical formulation are currently limited, there are many opportunities for SSE as adoption of the technology increases.

5.1 Bioprinting

3D bioprinting refers to the use of 3D printers to fabricate structures consisting of biological matter. The demand for engineered tissues is rising rapidly owing to the limited availability of donor tissues and organs for transplantation. SSE printing has therefore been adopted in this field due to its low printing temperature and semi-solid based feedstock, allowing high cell densities to be viably deposited. Many

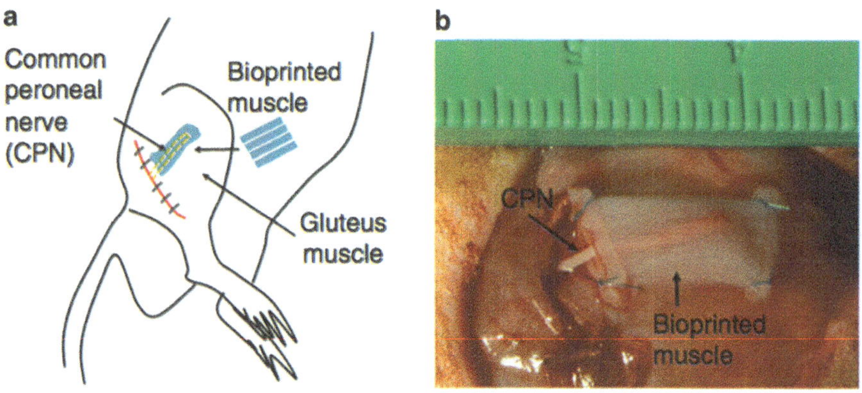

Fig. 7.9 (**a**) Schematic diagram of the rodent bioprinted skeletal muscle attachment (**b**) image of bioprinted muscle attachment. Reprinted with permission from [1]

studies have investigated the use of this technology in varying applications of bioprinting, the full scope of which can be found in a number of review articles [7, 10, 18–21]. This chapter will briefly outline some major recent advances, in order to fully demonstrate the extent to which SSE has been applied in the field.

In adapting SSE printers to form a novel integrated tissue–organ printer (ITOP), researchers have been able to extrude cell-laden hydrogels which can fabricate bone, cartilage and skeletal muscle, using clinical imaging data of an anatomical defect to shape the tissue construct [1]. Using a polymer framework as a support, hydrogels were used which contained gelatin, fibrinogen, hyaluronic acid (HA) and glycerol mixed into DMEM (high glucose) containing the cells required to create the desired tissue. The bone structures used amniotic fluid–derived stem cells which were able to develop into osteogenic lineages whilst a human ear structure used rabbit ear chondrocytes to create cartilage tissue. Finally, mouse myoblasts were incorporated into the gel to fabricate skeletal muscle tissue, able to mature into functional murine muscle *in vivo* (Fig. 7.9).

However, it is not always necessary to develop specific SSE printers in order to print tissues. Duan et al. used the Fab@Home Model 2 (Fig. 7.3) to print a hydrogel containing human aortic valvular interstitial cells to create a human heart valve (Fig. 7.1) [3]. The valve displayed high cell viability but it was not able to fulfil the full mechanical range of native valve tissue.

Perhaps one of the most interesting applications of the technology thus far has been to create bionic ears [22]. Using a Fab@Home Model 2, a team at Princeton University formulated a chondrocyte-seeded hydrogel matrix to create a cartilage ear structure within which an electrically conductive silver nanoparticle antenna was printed (Fig. 7.10). This antenna was connecting to cochlea-shaped electrodes supported on silicone, which not only enabled the ear to sense radio frequency reception, but also complementary left and right ears were able to process stereo audio music.

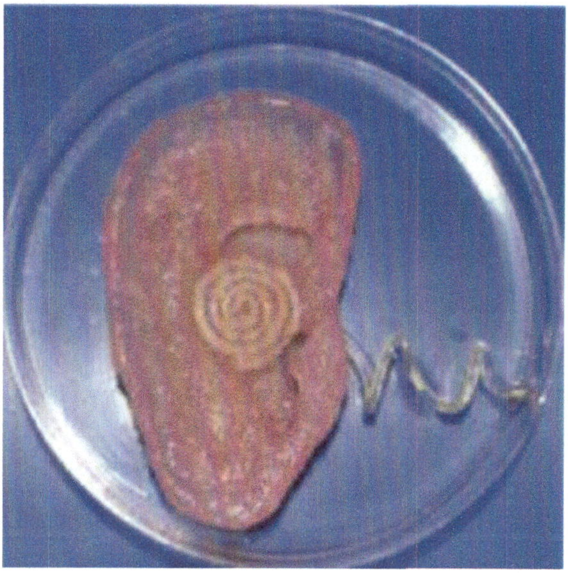

Fig. 7.10 3D printed bionic ear formed from cartilage and silver nanoparticle antenna. Reprinted with permission from [22]

5.2 Dental

3D printing has also been extensively applied to the field of dentistry from surgical models to replacement teeth, the full scope of which is beyond this chapter. Stratasys have recently introduced two specifically designed SSE printers for the purposes of dentistry known as CrownWorx™ and FrameWorx™. The printers extrude a form of wax designed to allow dental laboratories to create tailor-made crowns and bridges [23].

5.3 Confectionary

A novel opportunity may have been developed within the food industry in which a strand of SSE printers have utilised pectin-based gels to create chewable products as opposed to previous studies which use an HPMC base [24]. Katjes Magic Kandy Factory has created a form of SSE printer able to fabricate structures out of confectionary. The printer allows consumers to personalise their 'candy' by selecting from a pre-approved range of shapes and colours [24–26]. Of note however, is the technology's ability to operate without requiring a post-printing drying step, a result of the gel formulation being able to solidify sufficiently within seconds of printing. Due to the consistency of the feedstock, the gel does require heating (~70°C) and so may degrade some highly heat sensitive compounds [26]. This temperature however is only 30°C higher than the drying temperature required by current SSE printers. The viscosity necessitates an extrusion head with a larger orifice

(1 mm) than that of other SSE devices (0.4–0.8 mm) [8, 13, 16], thus reducing the resolution of the product created. Nevertheless, due to the low resolution of the extruder, the printing is significantly faster, and therefore is considered a more efficient manufacturer than most SSE systems.

6 FDA Regulation of Semi-Solid Extrusion

Much of the research conducted using semi-solid extrusion is very much within the proof-of-concept stages. As such, there is little need for any regulatory guidance regarding this technology, particularly within the field of pharmaceutical formulation. The only use of SSE printing that has warranted regulatory involvement by the FDA is the use of the technology in dentistry. The agency has approved the use of specific SSE feedstocks as dental devices which are therefore covered under the current good manufacturing practice (cGMP) requirements for devices (21 CFR part 820) as well as draft guidance issued for 3D printed medical devices [27, 28]. Approval means that the materials are considered 'finished devices that are suitable for use by health care professionals and are patient-matched or fitted at the point of care'. Examples of such materials include dental restorative devices such as direct filling resins, dental cements, denture resins, crowns and bridges. The regulation surrounding these devices is precise; each material needs to be approved to make a specific device with regards to physical properties and for its specific indication. In order to receive approval, the FDA requires an evaluation of the material's performance to demonstrate that it has the appropriate physical properties for its intended use.

7 Advantages of Semi-Solid Extrusion

Upon utilising semi-solid extrusion technology, an immediately apparent benefit is the simplicity of the drug loading process, as the API and excipients can be added directly to a gel base. This process allows the drug contents to be determined much more easily as it simply reflects the concentration of compound added, so long as it is mixed homogenously and not degraded during printing. It should be noted however that the composition of the gel can greatly affect its viscosity. By manipulating the concentrations of various components, important parameters such as the pH are also altered. This can play a significant role in the physical state of the material, and thus optimisation is a necessity to ensure the feedstock is able to print.

The technology has displayed a proficiency in creating complex formulations with the ability to be personalised not only by dose, but also release profiles. This however, is indicative of most 3D printing technologies and thus can be achieved by many other technologies discussed in this book. The real potential for SSE lies in both its low heating temperature and printing speed. Many pharmaceutically active drugs have a certain degree of thermolability, and thus, degrade at the operating temperatures of some 3D printers. The feedstock for SSE however requires much less heat to reach a viscosity able to print. By reducing the heat applied to the material it is therefore possible to utilise more thermolabile compounds. A consequence of this is the need for drying process post-printing to ensure the product has solidified sufficiently (discussed later). In terms of printing speed, SSE is able to reach rates of 20 mm/s [9] which is much higher than other forms of 3D printing. This efficiency is essential if the technology is to be implemented in a clinical setting as significant disruption to the working procedures of pharmacists or clinicians will almost definitely prevent wide-scale adoption.

8 Limitations of Semi-Solid Extrusion

When printing with SSE a clear limitation of applicability is the low resolution at which it is able to print. The nozzle heads of SSE printers must be able to accommodate a highly viscous material, and thus, are often much wider than the extrusion orifices of other printers. This low resolution can hinder the accuracy with which tablets are developed, potentially equating to an unreliable dose or dissolution profile. However, the low resolution also permits the technology to reach the high printing speed discussed earlier.

Due to the semi-solid nature of the feedstock, a form of post-printing drying is often required to ensure the 3D printed tablets have dried sufficiently to remain structurally viable and retain shape upon storage. There is a risk of the structure collapsing before sufficiently cooling, thus, a drying step may be essential in fabricating some less viscous pharmaceutical formulations. The Roberts group found that in order to ensure complete hardening had occurred, vacuum drying at 40°C was required for 24 hours [13, 16]. Interestingly, this was not required when creating the controlled-release bilayer tablets, suggesting that perhaps more viscous formulations do not require such a step [8].

9 Conclusion

Whilst 3D printing is changing the way we look at pharmaceutical formulation, the adoption of semi-solid extrusion is lagging and perhaps in the shadow of the more popular form of extrusion printing, FDM. Nevertheless, the research undertaken so far has demonstrated the ability of SSE to create complex multi-active tablets, personalising both the dose and release mechanism. Furthermore, the technology has facilitated some major advancements in the field of regenerative medicine, creating heart valves and bionic ears. Thus, by building on these applications, there is a great potential to maximise the abilities of SSE and make clinical adoption of 3D printing a real possibility.

References

1. Kang H-W, Lee SJ, Ko IK, Kengla C, Yoo JJ, Atala A. A 3D bioprinting system to produce human-scale tissue constructs with structural integrity. Nat Biotechnol. 2016;34(3):312–9.
2. Fussell S. This company is creating incredible 3D printed food you can eat. Business Insider. 2016. Available from: http://uk.businessinsider.com/3d-printed-food-foodini-2016-4?r=US&IR=T.
3. Duan B, Kapetanovic E, Hockaday LA, Butcher JT. Three-dimensional printed trileaflet valve conduits using biological hydrogels and human valve interstitial cells. Acta Biomater. 2014;10(5):1836–46.
4. Trenfield SJ, Awad A, Goyanes A, Gaisford S, Basit AW. 3D printing pharmaceuticals: drug development to frontline care. Trends Pharmacol Sci. 2018;39(5):440–51.
5. Awad A, Trenfield SJ, Goyanes A, Gaisford S, Basit AW. Reshaping drug development using 3D printing. Drug Discov Today. 2018; https://doi.org/10.1016/j.drudis.2018.05.025.
6. 3DDiscovery: RegenHU. 2017. Available from: https://www.regenhu.com/.
7. Critchley L. 3D Bio-printing approaches for tissue engineering applications; 2017. Available from: http://www.azom.com/article.aspx?ArticleID=13810.
8. Khaled SA, Burley JC, Alexander MR, Roberts CJ. Desktop 3D printing of controlled release pharmaceutical bilayer tablets. Int J Pharm. 2014;461(1–2):105–11.
9. Jeffrey L, Daniel C, Michael H, Maxim L, Warren P, Garrett B, et al. Fab@home model 2: towards ubiquitous personal fabrication devices. Ithaca: Cornell University; 2009.
10. Ozbolat I. 3D bioprinting. Academic Press. 2016;(1):93–124.
11. Prasad LK, Smyth H. 3D printing technologies for drug delivery: a review. Drug Dev Ind Pharm. 2015;42(7):1019–31.
12. Awad A, Trenfield SJ, Gaisford S, Basit AW. 3D printed medicines: A new branch of digital healthcare. Int J Pharm. 2018;548(1):586–96.
13. Khaled SA, Burley JC, Alexander MR, Yang J, Roberts CJ. 3D printing of tablets containing multiple drugs with defined release profiles. Int J Pharm. 2015;494(2):643–50.
14. Wald N, Law M. The Indian Polycap Study (TIPS). Lancet. 2009;374(9692):781. author reply 2
15. Rowe CW, Katstra WE, Palazzolo RD, Giritlioglu B, Teung P, Cima MJ. Multimechanism oral dosage forms fabricated by three dimensional printing. J Control Release. 2000;66(1):11–7.
16. Khaled SA, Burley JC, Alexander MR, Yang J, Roberts CJ. 3D printing of five-in-one dose combination polypill with defined immediate and sustained release profiles. J Control Release. 2015;217:308–14.
17. Li Q, Guan X, Cui M, Zhu Z, Chen K, Wen H, et al. Preparation and investigation of novel gastro-floating tablets with 3D extrusion-based printing. Int J Pharm. 2018;535(1–2):325–32.
18. Huang Y, Zhang XF, Gao G, Yonezawa T, Cui X. 3D bioprinting and the current applications in tissue engineering. Biotechnol J. 2017;12(8):1–16.
19. Jessop ZM, Al-Sabah A, Gardiner MD, Combellack E, Hawkins K, Whitaker IS. 3D bioprinting for reconstructive surgery: principles, applications and challenges. J Plast Reconstr Aesthet Surg. 2017;70(9):1155–70.
20. Murphy SV, Atala A. 3D bioprinting of tissues and organs. Nat Biotechnol. 2014;32(8):773–85.
21. Wang X, Ao Q, Tian X, Fan J, Wei Y, Hou W, et al. 3D bioprinting technologies for hard tissue and organ engineering. Materials (Basel). 2016;9(10):802.
22. Mannoor MS, Jiang Z, James T, Kong YL, Malatesta KA, Soboyejo WO, et al. 3D printed bionic ears. Nano Lett. 2013;13(6):2634–9.
23. Wyman C. New stratasys 3D printers expand digital dentistry with low cost, highly accurate wax-ups. Strasys Blog. 2014. Available from: http://blog.stratasys.com/2014/05/13/crownworx-frameworx-dental-3d-printers/.
24. Magic Candy Factory. Magic Candy Factory Katjes; 2017. Available from: http://magiccandyfactory.com.

25. Aydin B. Gel-comprising cartridge for the preparation of confectionery. Google Patents; 2015. EP3135117A1.
26. Aydin B, Fassin B, Ossenbrink D. Production of soft confectionery by means of additive process with a nozzle and soft confectionery obtainable thereby. 2016. WO2016113318A1.
27. Medical Devices Quality System Regulation. 2017. Available from: https://www.fda.gov/MedicalDevices/DeviceRegulationandGuidance/PostmarketRequirements/QualitySystemsRegulations/default.htm.
28. U.S. Food and Drug Administration. Technical considerations for additive manufactured devices. Draft guidance for industry and Food and Drug Administration staff. 2016. Available from: https://www.fda.gov/downloads/MedicalDevices/DeviceRegulationandGuidance/GuidanceDocuments/UCM499809.pdf.

Chapter 8
A New Dimension: 4D Printing Opportunities in Pharmaceutics

Jack Firth, Simon Gaisford, and Abdul W. Basit

Abstract 4D printing is a novel concept which adopts time as the fourth-dimension, allowing printed structures to independently alter their morphology. Using 'smart materials' as a feedstock, these products respond to stimuli such as light, temperature, water or pH, contorting in a specific manner to create structures able to adapt and respond to their environment. Such is the novelty of the technology, 4D printing is yet to be utilised in the field of pharmaceutical formulation. Nevertheless, smart materials have already been adopted in the industry, creating bioadhesive devices that respond to pH for specific targeting of areas of the GI tract, or even remotely-controlled 'micro-robots' designed to deliver chemo-therapeutics directly to the site of the tumour. Consequently, previous innovation with smart materials will likely act as precursors to the inevitable adoption of 4D printing in the industry. This will pave way to the creation of therapeutics that combine the personalisation afforded by additive manufacturing with the intelligent kinetics of the smart materials.

Keywords Four-dimensional printing · Three-dimensional printing · Bioprinting · Medical devices · Personalized medicines · Drug delivery systems

J. Firth (✉)
Department of Pharmaceutics, UCL School of Pharmacy,
University College London, London, UK

Intract Pharma Ltd., London, UK
e-mail: jack.firth@intractpharma.com

S. Gaisford · A. W. Basit
Department of Pharmaceutics, UCL School of Pharmacy,
University College London, London, UK

FabRx Ltd., Ashford, Kent, UK
e-mail: a.basit@ucl.ac.uk

A. W. Basit, S. Gaisford (eds.), *3D Printing of Pharmaceuticals*,
AAPS Advances in the Pharmaceutical Sciences Series 31,
https://doi.org/10.1007/978-3-319-90755-0_8

1 Introduction

Innovation does not rest. Though we are only beginning to explore the use of 3D printing in pharmaceuticals [1–5], 4D printing technology is already becoming a reality. Using time as the fourth-dimension, 4D printing can create products that independently alter their morphology in response to certain external stimuli. These changes in structure result from the use of smart materials, which respond to stimuli such as light, temperature, water or pH. Although 4D printing technology has not yet been used in pharmaceutical formulation, the use of smart materials has been widely researched, both to encapsulate specific drugs or cells and to also create bio-adhesive devices. Pharmaceutical formulation using smart materials can therefore, act as predictors of likely future applications for 4D printing. By combining 3D printing that is capable of creating personalised medicines and smart materials able to alter their own morphology, will 4D printing signal the dawn of 'smart' personalised formulations?

2 4D Printing

Building on the foundations of 3D printing, the 4D printing process incorporates a fourth-dimension other than the x, y and z axis; namely time. The products created are able to change configuration over time, often in response to external stimuli such as heat or water. Transformation into this 'fourth dimension' can be a result of the feedstock itself (using smart materials) and the predetermined design in which the product is created (known as smart design). This section will briefly describe the technology and applications used by 4D printers in a general sense; however more detailed reviews are available [6–8].

2.1 4D Printers

Research has initially been conducted on typical commercial 3D printers. However, when printing these constructs, particularly using smart materials, the feedstock can become agglomerated in conventional machines and as a result, some research groups have developed specific 4D printers. For instance, Choi et al. developed a printer with a coated extrusion nozzle, adapted for printing smart materials (in this case, thermal polyurethane, also known as TPU) via material extrusion 3DP [6]. Because TPU compresses when heated, the nozzle is easily clogged. To overcome this, the printer contains a heating bed to ensure sufficient heat circulation during the printing process, with the heating device placed adjacent to the nozzle to minimize heat loss. The nozzle itself is considerably longer than conventional machines and is coated in polytetrafluoroethylene to reduce friction. A further example of a specific 4D printer is that created by Ge et al. [9]. Based upon micro-SLA printers, the machine can create structures with a resolution of up to 1 μm. The printer utilises specific photo-curable smart materials which require a much higher exposure energy than typical acrylate-based polymers and thus would not be suitable for printing using conventional SLA.

2.2 *Smart Materials*

Smart materials, in relation to 4D printing, are feedstocks that have the ability to alter their shape in response to certain stimuli. There are two main types of smart material that have been used in 4D printing thus far; hydrogels which swell when exposed to water or other solutes and shape memory polymers (SMPs). These SMPs respond to a range of stimuli, including temperature, pH or UV radiation.

Hydrogels are used to create structures in which the diffusion of water into the polymer network creates specific, localised swelling, thus manipulating the morphology of the product. By printing a rigid waterproof material connected by a hydrogel at 'hinges', researchers at MIT created a linear product that, upon immersion in water, contorted to create a 3D structure. The absorption of water into the hinge material causes swelling, which leads to predicted bending of the rigid material. This was used to create both a fully enclosed cube as well as a cube outline dependent upon the size and manipulation of the structural, waterproof sections [10].

Conversely, SMPs are capable of adopting temporary morphologies in response to certain external stimuli such as temperature or pH which then recover to the permanent shape upon returning to the original external environment. In mechanisms such as this, the smart material used responds to a certain predicted stimulus reaching a critical inflection point such as the glass transition temperature (Tg) of a polymer. At the rested permanent form, the temperature of the polymer should be below the Tg, ensuring the material is in the brittle, glass state. However, at temperatures above the Tg, the material will transition into a viscous, more flexible state. This allows for specific, predicted alterations in the structure morphology, before returning to the rested state when the temperature falls below Tg.

One such application of these SMPs has been to create stents able to contort in response to the changing temperature when inserted into a patient (Fig. 8.1). Due to the complex geometries required by stents, conventional fabrication methods are difficult and time-consuming. However, by adopting 4D printing it is possible to create stents of any size in a time and cost-efficient manner. Using SMPs, a cardiovascular stent has been created that possesses a small temporary structure at a heated state which would allow for insertion with a minimally invasive procedure [9]. Following its implantation, the stent then cools and returns to its larger, clinically effective shape where it permanently remains. Not only has this mechanism been applied to tracheal stents allowing for individual patient personalisation, it has also been developed to create 'smart' stents [11, 12]. These 'smart' structures have iron oxide incorporated into a thermo-responsive PLA-based ink which allows them to be guided magnetically and stimulated remotely.

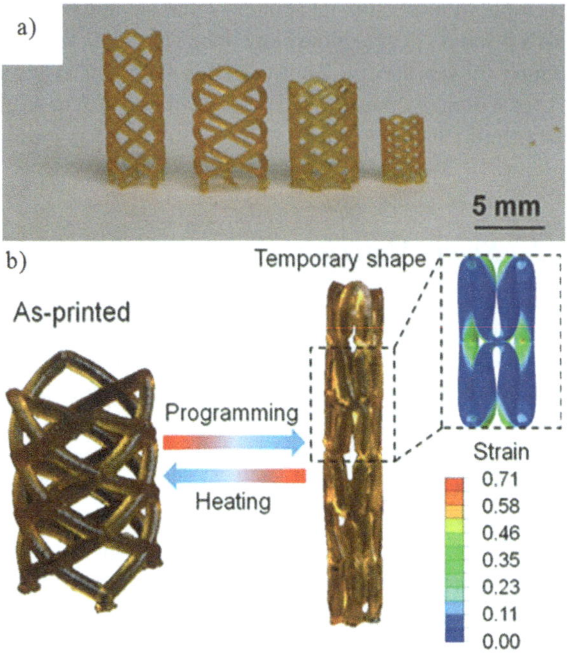

Fig. 8.1 (**a**) Various printed cardiovascular stents (**b**) Change in stent morphology by altering temperature. (Reprinted with permission from [9])

2.3 Smart Design

In addition to smart materials, a method of creating 4D printed structures is through the design of the product itself. By manipulating the design of the structure and the orientation of smart materials, it is possible to induce specific morphological changes in the end-product when exposed to certain stimuli. For example, Ge et al. created a structure with SMP fibres deposited at a prescribed orientation [9]. Following heat exposure, the printed structure transforms into complex shapes with spatially varying curvatures depending upon the orientation in which the fibres were initially designed. Therefore, by combining both smart materials and smart design, 4D printing can create structures with complex morphological changes.

Through a combination of smart materials and smart design, 4D printing can facilitate the fabrication of structures too complex to be obtained via 3D printing. By printing an initial design that is very simple in its morphology, the process is improved both in terms of speed and cost-efficiency whilst the logistics of storing such objects is made much more convenient. Furthermore, by initially printing 2D structures, much of the complications during 3D printing can be avoided as there is less risk of the feedstock clogging or the structure collapsing before solidifying. Then when required, the structure can take the necessary 3D structure to enact its intended effect.

3 Pharmaceutical Opportunities

Due to its novelty, 4D printing as a concept has yet to be applied to pharmaceutical formulation. Not only has the concept just begun to move from proof-of-principle to the search for real clinical applications, there is still much research needed before the technology can be applied to consumable medications. That being said, 4D printing essentially consists of using 3D printers to fabricate structures made from smart materials able to self-fold or unfold. Though not created with 3D printing, these smart materials have previously been applied to pharmaceutics and drug delivery. Therefore, this research can be used to predict the likely applications for the technology once it moves into the pharmaceutical arena. The use of these smart materials can be grouped into two mechanisms based upon the method of drug delivery applied; bio-adhesion and encapsulation.

3.1 Bio-adhesion Devices

Bio-adhesion devices are able to attach to the intestinal endothelium whereby they can initiate drug release. He et al. investigated this application developing a tri-layered, muco-adhesive drug delivery system designed to act as a form of sustained release formulation (Fig. 8.2) [13]. The device used a pH sensitive hydrogel as an outer layer, which upon reaching a pH of 6.5 in the small intestine, would contort into a shape able to grip onto the gut wall. This was successfully demonstrated to adhere to the walls of a porcine small intestine (using a pH 6.5 buffer), providing a longer residence time than a standard poly(ε-caprolactone) patch (103 and 72 min respectively) and minimising the drug exposure to intestinal fluid. Using this self-folding mechanism, the device can ensure that the drug-loaded muco-adhesive layer affixes to the intestinal wall for longer, thus allowing more drug to traverse the mucosal epithelium.

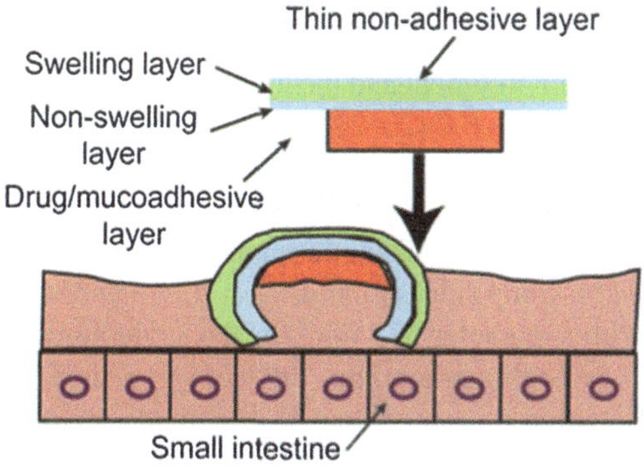

Fig. 8.2 Tri-layered, muco-adhesive drug delivery system. (Reprinted with permission from [13, 14])

A further bio-adhesive formulation using smart-materials are thermo-responsive multi-fingered drug eluting devices, otherwise known as 'theragrippers'[14]. Adopting its predetermined design when exposed to temperatures above 32°C, this device spontaneously grips onto tissue when introduced into the body from room temperature. The porosity of the structures allows drug loading which is then released for up to 7 days with first order release kinetics. This was demonstrated by in vitro studies in which the theragrippers displayed an enhanced sustained release of doxorubicin when compared with a control.

Building upon this concept, some researchers have incorporated iron oxide nanoparticles into the porous hydrogel layer [15]. In doing so, the micro-grippers become magnetically responsive and can therefore be guided remotely. By using a magnetic guide wire, the micro-grippers could potentially be directed to their target site within a clinical setting or during surgery. In some cases, the devices were powerful enough to excise cells from a fibroblast cell clump, suggesting their potential use in actually enacting some form of surgery themselves.

3.2 Encapsulation Devices

Encapsulation devices are self-folding structures that alter their shape to create a container in which pharmaceutical compounds or even cells can be enclosed (Fig. 8.3a) [17]. Fibroblasts and pancreatic beta cells have been encapsulated and displayed viability for over a week post-encapsulation (Fig. 8.3b, c) [16]. Yeast cells have also been contained in such devices using self-folding capsules that were able to release said contents dependent upon the surrounding temperature [18]. This technology is thought to be well-suited as a form of enteric drug delivery device whereby the content is released upon certain stimuli, be it pH or time.

It is not only cells that have been utilised in these form of encapsulation devices, as anticancer compounds have also been investigated. Li et al., created a 'micro-robot' device consisting of a hydrogel bilayer fabricated by conventional lithography [19]. A pH-responsive gel acted as one layer, able to change the morphology of the structure to trap and release a cargo, whilst the other layer contained iron oxide particles, enabling the device to be guided magnetically (similar to the micro-grippers discussed earlier). Resulting from the high metabolic rate of the tumour microenvironment and the low oxygen partial pressure, tumour tissue pH is approximately 4.5–6.0 [20–22]. This is utilised by the micro-robot which is designed to release its cargo after encountering acidic tumour tissue. Microbeads loaded with the anti-cancer drug PCL-DTX were encapsulated by the device and transported to cells using the magnetic nature of the outer layer. The structure was shown to successfully release the drug at the intended site, reducing cell viability by 70%. With a diameter of 3 mm, the device requires significant reduction in size in order to pass throughout the circulatory system, and therefore be clinically viable. However, this technology does provide proof-of-concept evidence that SMPs have the ability to target anti-cancer drug delivery. By focusing release at the tumour

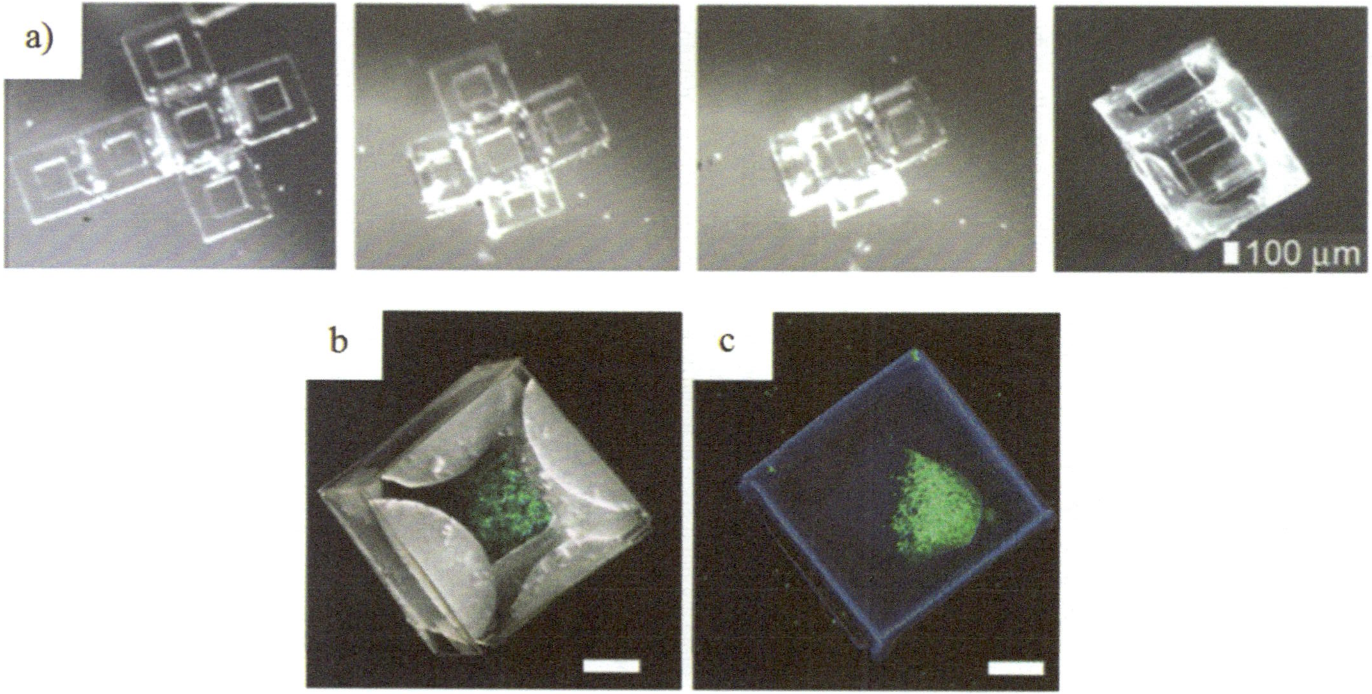

Fig. 8.3 (a) Images showing the self-folding process of a polymeric container at 60°C (b) bright-field and (c) fluorescence z-plane stack image of a non-porous polymeric container containing stained fibroblast cells. The scale bar is 250 µm long in Fig. 8.3 parts b & c. (Reprinted with permission from [16])

tissue, the therapeutic effect of the drug is maximised whilst limiting the amount of drug that passes throughout the systemic circulation. Therefore, the device could not only increase the efficacy of the drug, but also reduce the risk of serious side-effects which often limit the use of many anti-cancer compounds.

3.3 Bioprinting

Although not a direct opportunity within the field of pharmaceutical formulation, a significant application of 4D printing worth discussing is that of bioprinting. Discussed multiple times throughout this text, bioprinting is currently being used extensively in 3D printing [23, 24]. By transferring this knowledge into 4D printing, it is possible to not only create living structures, but ones that can respond to their environment.

As a form of smart material, hydrogels have been utilised by those in 4D printing primarily for their osmotic properties. However, the porosity that allows these materials to contort also makes them well-suited for bioprinting as it facilitates a controlled gas and nutritional supply to the cells they support [25]. Such hydrogels have in fact been utilised to create scaffold-free bio-printed structures, enabling a more efficient printing process. With the ability to facilitate native-like tissue growth, 4D printing can create self-assembling cell-based structures without any need for a scaffold, mould support or liquid delivery medium. This was demonstrated in one study whereby chondrocytes were deposited inside a form of hydrogel cylinder [26]. Upon printing, the hydrogel enabled the latticed strands of tissue to fuse together over time, creating a patch of viable tissue for implantation. Furthermore, due to the proliferating nature of the cells deposited in bioprinting, it is possible for this form of biological feedstock (called bioink) to act as the smart material itself. By using bioinks of different sizes and shapes, it has been proved possible to create scaffold-free mini-tissues that over time can integrate and develop into larger tissues [27]. In developing structures such as this, 4D printing is advancing towards a stage whereby a selection of bioinks could be printed which integrate into functioning, complex organs consisting of many tissue and cell types. This technology indicates that it may now be possible to use 4D printing to fabricate mature tissues, able to respond to external and internal stimuli, similar to that of real physiological organs. To that end, by using smart hydrogels that are responsive to specific ion levels, researchers have created tissues that are able to display muscle-like movement [26]. The cells in this tissue contracted in response to an influx of calcium ions, much like the physiological mechanism that drives normal muscle function. This demonstrates that 4D printing provides a significant opportunity to improve upon the bioprinting performed by 3D printing, creating complex structures that are able to respond to their environment. Although much work is required before this goal is achieved, research completed thus far demonstrates the unique ability of the technology to create stimuli-responsive structures in which the cells can mature.

4 Conclusion

By assessing the use of smart materials currently applied in pharmaceutical formulation, it is not hard to visualise the initial applications for 4D printing. Specific 4D printers based on micro-SLA machines are able to print smart materials at high resolutions and thus should be able to fabricate structures with the same complexity as the theragrippers for example. In applying the 4D printing process to such structures, not only would the process improve in terms of time and cost efficiency, but there would also be the opportunity for patient personalisation as described extensively throughout this book. For pharmaceutics, the most valuable application of this technology is the ability to create targeted bio-adhesion devices. Using pH as a stimulus to affix to specific regions of the gastrointestinal tract, these formulations can target drug release to the most beneficial site for the treatment or disease in question.

Though 3D printing is still in its infancy within the pharmaceutical sector, the transition to 4D seems a relatively simple one. By using smart materials as a feedstock, it is possible to create drug delivery technologies that can alter and change in response to the varying conditions within the gastrointestinal tract. In doing so, these 'smart drugs' can provide more targeted drug release that can be personalised for the physiological environment in each individual patient. 3D printing still requires much development before it can be clinically viable, however it would be wise to look to the future as 4D printing is likely to materialise sooner than one may think.

References

1. Trenfield SJ, Awad A, Goyanes A, Gaisford S, Basit AW. 3D printing pharmaceuticals: drug development to frontline care. Trends Pharmacol Sci. 2018;39(5):440–5.
2. Awad A, Trenfield SJ, Gaisford S, Basit AW. 3D printed medicines: A new branch of digital healthcare. Int J Pharm. 2018;548(1):586–96.
3. Alhnan MA, Okwuosa TC, Sadia M, Wan KW, Ahmed W, Arafat B. Emergence of 3D printed dosage forms: opportunities and challenges. Pharm Res. 2016;33(8):1817–32.
4. Zema L, Melocchi A, Maroni A, Gazzaniga A. Three-dimensional printing of medicinal products and the challenge of personalized therapy. J Pharm Sci. 2017;106(7):1697–705.
5. Awad A, Trenfield SJ, Goyanes A, Gaisford S, Basit AW. Reshaping drug development using 3D printing. Drug Discov Today. 2018; https://doi.org/10.1016/j.drudis.2018.05.025.
6. Choi J, Kwon O-C, Jo W, Ju Lee H, Moon M-W. 4D printing technology: a review. 3D Print Addit Manuf. 2015;2(4):159–67.
7. Farhang M, Seyed MMHN, Xun L, Jun N. A review of 4D printing. Mater Des. 2017;122:42–79.
8. Gladman AS, Matsumoto EA, Nuzzo RG, Mahadevan L, Lewis JA. Biomimetic 4D printing. Nat Mater. 2016;15(4):413–8.
9. Ge Q, Sakhaei AH, Lee H, Dunn CK, Fang NX, Dunn ML. Multimaterial 4D printing with tailorable shape memory polymers. Sci Rep. 2016;6:31110.
10. Tibbits S. 4D printing: multi-material shape change. Archit Des. 2014;84:116–21.

11. Wei H, Zhang Q, Yao Y, Liu L, Liu Y, Leng J. Direct-write fabrication of 4D active shape-changing structures based on a shape memory polymer and its nanocomposite. ACS Appl Mater Interfaces. 2017;9(1):876–83.
12. Zarek M, Mansour N, Shapira S, Cohn D. 4D printing of shape memory-based personalized endoluminal medical devices. Macromol Rapid Commun. 2017;38(2):1600628.
13. He H, Guan J, Lee JL. An oral delivery device based on self-folding hydrogels. J Control Release. 2006;110(2):339–46.
14. Malachowski K, Breger J, Kwag HR, Wang MO, Fisher JP, Selaru FM, et al. Stimuli-responsive theragrippers for chemomechanical controlled release. Angew Chem Int Ed Engl. 2014;53(31):8045–9.
15. Breger JC, Yoon C, Xiao R, Kwag HR, Wang MO, Fisher JP, et al. Self-folding thermo-magnetically responsive soft microgrippers. ACS Appl Mater Interfaces. 2015;7(5):3398–405.
16. Azam A, Laflin KE, Jamal M, Fernandes R, Gracias DH. Self-folding micropatterned polymeric containers. Biomed Microdevices. 2011;13(1):51–8.
17. Yoon C, Xiao R, Park J, Cha J, Nguyen T, Gracias D. Functional stimuli responsive hydrogel devices by self-folding. Smart Mater Struct. 2014;23:094008.
18. Stoychev G, Puretskiy N, Ionov L. Self-folding all-polymer thermoresponsive microcapsules. Soft Matter. 2011;7:3277–9.
19. Li H, Go G, Ko SY, Park J-O, Park S. Magnetic actuated pH-responsive hydrogel-based soft micro-robot for targeted drug delivery. Smart Mater Struct. 2016;25(2):027001.
20. Manchun S, Dass CR, Sriamornsak P. Targeted therapy for cancer using pH-responsive nanocarrier systems. Life Sci. 2012;90(11–12):381–7.
21. Swietach P, Vaughan-Jones RD, Harris AL, Hulikova A. The chemistry, physiology and pathology of pH in cancer. Phil Trans R Soc B. 2014;369(1638):20130099.
22. Tong ZQ, Luo WH, Wang YQ, Yang F, Han Y, Li H, et al. Tumor tissue-derived formaldehyde and acidic microenvironment synergistically induce bone Cancer pain. PLoS One. 2010;5(4):e10234.
23. Peng W, Datta P, Ayan B, Ozbolat V, Sosnoski D, Ozbolat IT. 3D bioprinting for drug discovery and development in pharmaceutics. Acta Biomater. 2017;57:26–46.
24. Goole J, Amighi K. 3D printing in pharmaceutics: a new tool for designing customized drug delivery systems. Int J Pharm. 2016;499(1–2):376–94.
25. Kalaskar D, Das A. Four dimensional printing in healthcare: Woodhead Publishing; 2007. p. 207–18.
26. Yu Y, Moncal KK, Li J, Peng W, Rivero I, Martin JA, et al. Three-dimensional bioprinting using self-assembling scalable scaffold-free “tissue strands” as a new bioink. Sci Rep. 2016;6:245–56.
27. Mironov V, Visconti RP, Kasyanov V, Forgacs G, Drake CJ, Markwald RR. Organ printing: tissue spheroids as building blocks. Biomaterials. 2009;30(12):2164–74.

Chapter 9
Medical Applications of 3D Printing

Grace B. Hatton, Christine M. Madla, Simon Gaisford, and Abdul W. Basit

Abstract Three-dimensional printing (3DP) is an entirely novel method of manufacture with its applications only limited by the imagination. The mainstay of 3DP utilisation practically to date has been in the field of engineering, largely for the purpose of generating model prototypes. However, the potential of 3DP has increasingly been recognised in areas of commercial manufacture in medicine due to its capacity to produce materials and devices that can equal, if not surpass, the benefits of traditional consumer goods. The opportunities for future uses are innumerable ranging from tissue engineering, the on-demand fabrication of medical devices and advanced applications in other fields with the same pressing need for medical personalisation. The 3D printing arena is ultimately exciting and endless in opportunities with the FDA encouraging the development of science and risk based approaches. This chapter will discuss the existing and future medical applications of 3DP and its potential to revolutionise medical practice.

Keywords 3D printing · Bioprinting · Tissue engineering · Medical devices · Personalized medicines · Drug delivery systems

G. B. Hatton (✉) · C. M. Madla
Department of Pharmaceutics, UCL School of Pharmacy,
University College London, London, UK
e-mail: grace.hatton@nhs.net

S. Gaisford · A. W. Basit
Department of Pharmaceutics, UCL School of Pharmacy,
University College London, London, UK

FabRx Ltd., Ashford, Kent, UK

A. W. Basit, S. Gaisford (eds.), *3D Printing of Pharmaceuticals*,
AAPS Advances in the Pharmaceutical Sciences Series 31,
https://doi.org/10.1007/978-3-319-90755-0_9

1 Introduction

As extensively discussed in previous chapters already, the advent of three-dimensional printing (3DP) is a revolution in itself. The revolutionary aspect of this technology is two-fold; not only does the very principle offer an entirely novel method of manufacture, but its applications are limited only by the imagination. In healthcare alone, 3DP offers an advantage limited by other approaches: Personalisation. Prosthetic devices (such implantable defibrillators, medicines and equipment) and even tissues and organs can be made-to-measure and made-to-order for a specific purpose, be that man or machine. The added benefits – cost-effectiveness; simplified production techniques; and increased opportunities for collaboration [1–3] – are equally attractive. This chapter will discuss the existing and future medical applications of 3DP and its potential to revolutionise manufacturing for this purpose.

The mainstay of 3DP utilisation practically to date has been in the field of engineering largely for the purpose of generating model prototypes. From architecture to vehicle design, 3DP has found its way as a viable method of facilitating rapid prototype production, modification and re-production owing to its ease of use and malleability. And yet, the potential of 3DP has increasingly been recognised in areas of commercial manufacture; indeed, 3DP has the capacity to produce materials and devices that can equal (if not better, thanks to the benefit of customisation) traditional consumer goods.

2 Materials for 3D Printing in Medicine

Prior to fabricating the final product, materials must be developed to ensure that the 3D printed product is bio-inert and/or biodegradable if its intended use is an implantation device, artificial organ or as prostheses. Materials employed for the purpose of medicinal use are described as "biomaterials", defined as nonviable materials used at the interface of biological entities [4]. Table 9.1 outlines the materials employed by different 3D printing technologies and their applications in the clinical field.

These materials are required to be completely biocompatible meaning that its desired function is successfully performed without any undesirable effects to the recipient. For example, materials for long-term implantation should be wear and corrosion resistant. Degradable and bioactive materials include a portfolio of polymers, ceramics and hydrogels used in tissue engineering and drug delivery applications, briefly discussed in the respective 3D printing technologies chapters.

Table 9.1 Biomaterials of choice for clinical use, adapted from [5]

Material	Current explored application
Polyester textile	Vascular grafts and heart stents
Polyurethane	Pacemaker lead insulation
Silicones	Soft tissue augmentation, insulating leads and ophthalmological devices
Poly(methyl methacrylate) (PMMA)	Bone cement and intraocular lenses
Polylactic acid (PLA)	Heart model studied for human transplantation
Carbon	Heart valves
Calcium phosphates	Bioactive surfaces and bone substitutes
Aluminium	Bearing surfaces
Stainless steel	Stents and orthopaedic implants
Nickel titanium naval ordnance laboratory (NiTiNol) composition	Shape memory applications (see Chap. 8)
Platinum group alloys	Electrodes
Cobalt-chromium alloys	Bearing surfaces, heart valves, stents and pacemaker leads
Titanium alloys	Dental implants, femoral stems, pacemaker cans, heart valves, fracture plates and spinal cages

2.1 Metals

The most common implantation device material are metals and their alloys due to their high strength and ductility, often utilised to replace damaged or diseased tissue in bearing applications where long-term fixture is required [6]. Titanium, cobalt-chromium and platinum alloys have been processed via powder based fusion technologies such as selective laser melting (SLM, see Chap. 5). There are different characteristics to consider when employing a metal biomaterial for the use of 3D printing. These can be classified as conventional metals and their alloys; shape memory alloys and biodegradable metals which will be discussed in turn.

2.1.1 Conventional Metals and their Alloys as Biomaterials

Conventional metals are described as bio-inert materials. As previously stated, titanium and stainless-steel alloys are often used for the manufacture of long-term implantation devices. Titanium is often the metal of choice when designing dental implants, knee replacements, hip implants (Fig. 9.1), heart valves and spinal cages. Stainless steels are low in cost and easily malleable, exhibiting strong advantages for its application in the medical field. However, challenges in processing stainless steel powder have deterred the mass-production of this biomaterial.

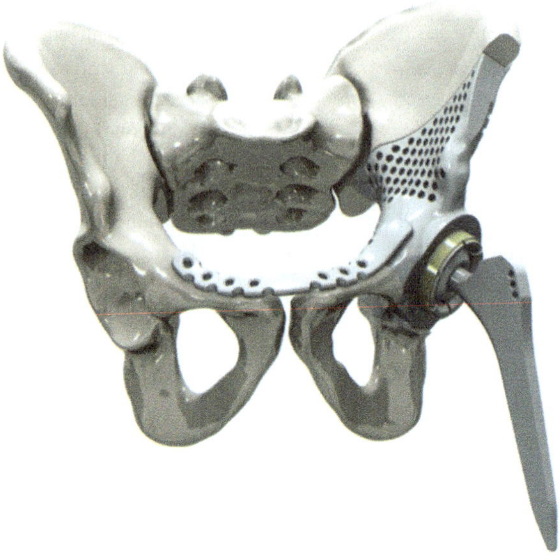

Fig. 9.1 CAD model of a titanium hip implant, fabricated with the use of SLM. (Reprinted with permission from [7])

2.1.2 Shape Memory Alloys

Shape memory allows (SMAs) (further discussed in Chap. 8) are biomaterials that, when subjected to external stimuli such as heat, pH, light or magnetic field, can undergo pseudoelastic deformations and recover to their original design following another memory stimuli. The most renowned SMA is the Nickel Titanium Naval Ordnance Laboratory (NiTiNol) composition. Its atomic structure allows the material to be present in the two different phases, providing a change in temperature, and was initially developed for large bone trauma. The use of SLM, however, has allowed the fabrication of NiTiNol as self-expanding cardiovascular stents (Fig. 9.2).

2.1.3 Biodegradable Metals

These metals are designed to degrade in the body after their designed function is complete in the form of an implant to support tissue growth for cardiovascular, intestinal and orthopaedic use.

A significant advantage using biodegradable metals is the reduction of surgical procedures; biodegradable implants are not required to be removed in an invasive secondary procedure following tissue healing, and thus, are often used in paediatric and geriatric patients in particular.

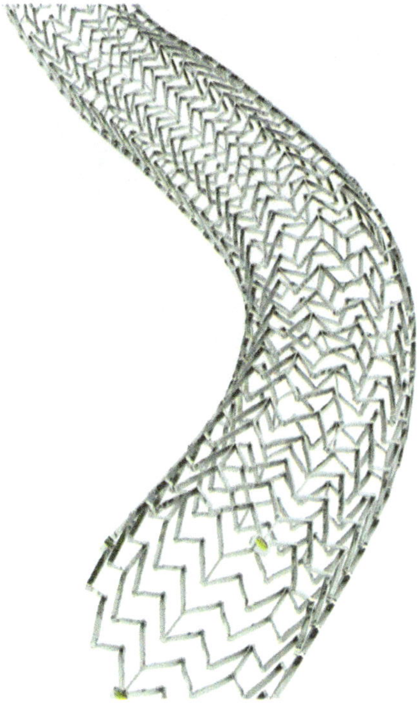

Fig. 9.2 An example of a drug-eluting cardiovascular stent fabricated using SLM with NiTiNol as the biomaterial. (Reprinted with permission from [8])

2.2 *Bio-Ceramics and Bioactive Glass*

Bioceramics and bioactive glass are biomaterials of major interest within the scope of orthopaedics due to their capabilities of fabricating patient specific implants. There are two types of bio-ceramics and bioactive glass namely, degradable and non-degradable.

2.2.1 Biodegradable and Bioactive Ceramics and Glass

In 1969, Larry Hench invented the first material to successfully form a chemical bondwith human bone (45S5®). This discovery introduced a new genre of materials, now known as bioactive glass [9]. 45S5® is a silica-based glass composed with sodium, calcium and phosphorous oxides. Synthetic hydroxyapatite is the most comparable chemical structure to the inorganic hydroxycarbonate apatite phase found in human bone tissue and is used as a dependable bone substitute. When repairing bone defects in humans, powder bed fusion technologies can be employed to 3D print personalised bone structures using bioactive glass (Fig. 9.3).

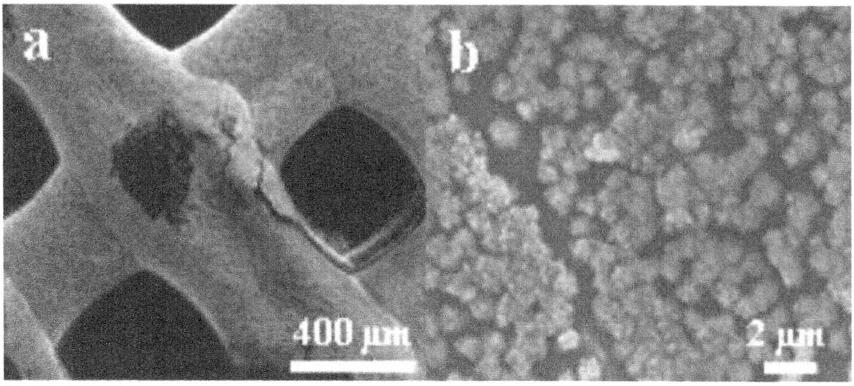

Fig. 9.3 SEM images of (a) surface of the bioactive glass scaffold printed without a hydrogel following immersion in an α-MEM for 14 days and (b) a magnified image showing the crystal formation on the surface. (Reprinted with permission from [10])

2.2.2 Non-degradable Bio-Ceramics

Like bioactive glass, bio-ceramics are often used in the clinical field for joint and bone substitutions. Traditional ceramics are bio-inert and, when coupled with metal alloys, are extremely hard, have high compressive strengths and low coefficients of friction [11]. One of the most notable developments in bio-ceramics was the use of alumina in joint surgeries, especially for ball and socket hip replacements. 3D printing technologies in this sector have concentrated on creating porous structures and have been limited to developing entire biomedical models solely utilising bio-ceramics.

2.3 Hydrogels

Hydrogels, as the name suggests, are organic soft solids that contain > 95% water [12] that are the starting materials employed when using bioprinting and for the 3D printing of pharmaceuticals [13–15]. Due to their composition, hydrogels are able to support biomolecules, proteins and cells that can be fabricated for tissue engineering purposes. The bioinks used for 3D bioprinting are available as natural and synthetic hydrogels successfully used within inkjet printing, semisolid extrusion printing and laser assisted printing. As 3D printed hydrogels are intended for use in the body, the bioinks are required to be biocompatible and should not elicit a negative interaction, provide strength for tissues that require mechanical support, gel at physiologically benign conditions and should retain its shape following printing.

Natural biopolymers are available as gelatine, chitosan and alginate which have been employed for tissue engineering and regenerative medicine. The most notable synthetic polymer currently used as a bioink is poly(ethylene glycol) (PEG) which presents great advantages over biopolymers. For example, PEG produces consistent batch products and demonstrates good hydrophilicity and protein repulsiveness. In addition, PEG is a Generally Recognised as Safe (GRAS) pharmaceutical excipient, approved by the Food and Drug Administration (FDA) [16].

2.4 Polymers

Since their approval for human use from the FDA, polymer science has extensively evolved to cater to a host of different applications in medical sciences. 3D printing technologies (semisolid extrusion printing (SSE), selective laser sintering (SLS) and stereolithography (SLA), in particular) have successfully fabricated porous scaffolds with a multitude of properties by virtue of its CAD model with polylactic acid, polycaprolactone and PEEK to name a few [17].

3 Applications of 3D Printing in Medicine

As it stands currently, 3D printing technologies are also being explored as a viable method of personalising medicines at the point of use and with a view to expand into rapid throughput screening of new drug candidates on 3D printing biological tissue to identify intra-individual therapeutic responses. Medicines production is one of the few fields in which the utilisation of 3D printing has lagged behind and yet has a great deal of potential. Indeed, 3D printing so far appears to have a proverbial finger in every pie from construction to computing, healthcare, aerospace medicine and even reproducing human organs for surgical transplantation using organic tissues as the printing material. The arguable limit of 3D printing utilisation is imagination and the resolution of the printer also plays a crucial role. Figure 9.4 illustrates the current application of 3D printing in medicine. Albeit the main challenge facing the application of 3D printing in a larger scale is its implications on regulatory standards related to quality assurance of the manufacturing process.

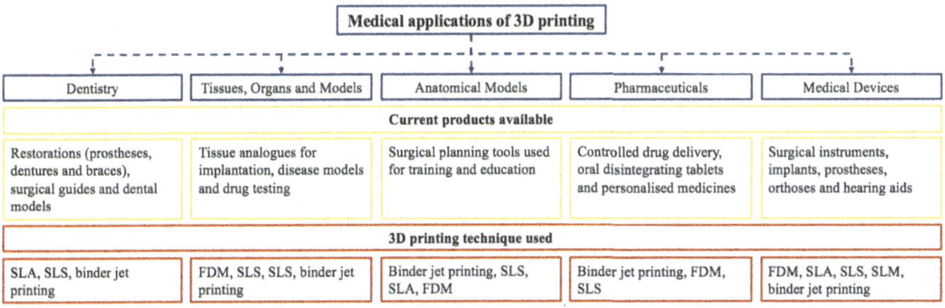

Fig. 9.4 Current medical applications of 3D printing in medicine

3D printing is competitive for small-scale production of products that require customisation and frequent modification, and for products that require complex geometries. As previously discussed, materials used to fabricate 3D printed products are meticulously assessed and employed in accordance to their final purpose. Various constructs have already been prepared using 3D printing in the medical field ranging from implants, cardiovascular stents, blood vessels [18], bone repair [19], cells and even 3D printed drugs (see Chap. 3) [20, 21].

The current medical uses for 3D printing can be characterised to three main categories; tissue and organ fabrication; creating patient specific medical devices (such as prosthetics and implants) and; anatomical 3D models used for surgical trainings, all of which will be discussed.

3.1 Bioprinting Tissues and Organs

More extensively discussed in Chaps. 7 and 8 respectively, there is an increasing demand for the bioprinting of tissues and organs. Advances in 3DP technology have broached the realm of regenerative medicine, ensuring that the printing of biological materials is now very much reality over fantasy. Indeed, such bioprinters are capable of printing not only stem cells but of building organs and blood vessels in a cell-by-cell fashion, enabling printing of tissues fit for human use on demand using automated, laser-calibrated print heads [22].

It is estimated that around twenty patients in the U.S.A alone die each day whilst awaiting organ transplantation [23], and though still premature as an option for addressing global organ donor shortages, 3DP offers a potential solution nonetheless [1, 24]. In this way, tissues and organs can potentially be printed using autologous stem cells in a similar vein of thinking to existing autologous approaches in medical management such as bone marrow transplantation. Such capacity would not only arguably remove the need for cadaveric or live-donor transplants (often at risk of rejection owing to tissue or cellular incompatibly with the recipient host alongside prolonged waiting lists for human "matches") but would also potentially allow for elective transplantation of organs in areas such as ageing and regenerative medicine which are both relatively new fields of investigation for pre-emptive treatment *per se* [25].

The U.S.A. stem cell research company Celprogen Inc.® is one such pioneer of 3D organ bioprinting having successfully engineered one of the world's first 3D printed human heart that is currently being validated for human use [26]. This was made from polylactic acid (PLA) (see Table 9.1) material that was populated with adult human cardiac stem cells. In December 2016, Celprogen Inc.® also announced the successful 3D printing of a human pancreas from PLA seeded with adult human pancreatic stem cells. The organ was coated with extracellular matrix protein and seeded with pancreatic stem cells from two different human lines which then successfully differentiated into a functional adult pancreas [27]. Similar to the work of Celprogen Inc.®, ETH Zurich have manufactured a silicone heart that is

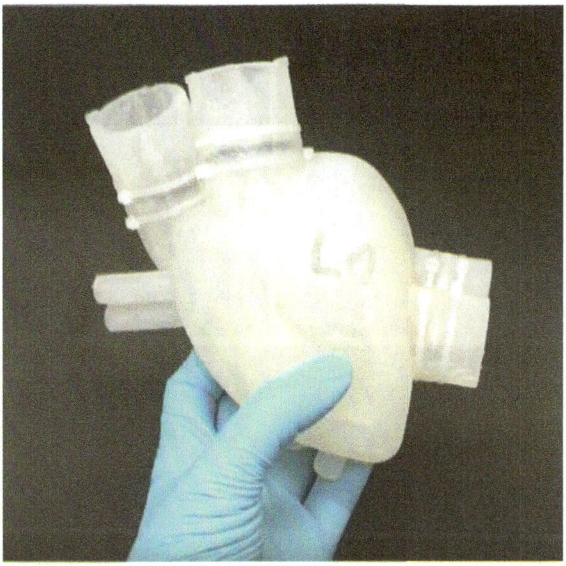

Fig. 9.5 3D printed silicone heart designed by ETH Zurich. (Reprinted with permission from [28])

capable of beating like the real organ using a lost-wax casting 3D printing technique (Fig. 9.5) [28].

Much work, however, is required to optimise the silicone 3D printed heart as its current iteration only lasts for 3000 beats, sufficient to keep someone alive for 30–45 min. Having said this, the 3D printing technique provides a promising solution to the lack of organ donations.

Equally, 3DP manufacture builds on more conventional approaches to tissue regeneration that had previously offered only simple "scaffolding" approaches to cellular tissue engineering [25, 29, 30]. Indeed, 3D printing offers precision-controlled cellular fabrication, and directly produces personalised tissues to predetermined dimensions. A number of researchers have already exploited these principles and successfully printed human cartilage [30], liver and tracheal splints for paediatric transplantation [29]. Others have expanded into the fields of orthopaedics (exoskeletons) and ophthalmology (corneal regeneration) [3, 19, 29, 31, 32].

As it stands today, it is still a significant challenge to recreate a fully functional organ for the purpose of transplantation with the adequate thickness and complex vascular network [4]. However, there is a larger portfolio of examples of the 3D printing of homogenous tissues without vasculature such as skin, bones and cartilage *in vitro,* as previously mentioned. A study by Mannoor et al. [33] have ventured into the fabrication of a fully functioning ear by the co-printing of nano-sized electronic elements with living cells in a hydrogel solution. The final product was a bionic ear with enhanced auditory sensing for radio frequency reception, printed in an anatomically precise manner (Fig. 9.6).

As thin living tissues, skin, cartilage, cell laden scaffold and blood vessels have been investigated and have successfully been fabricated by virtue of 3D bioprinting,

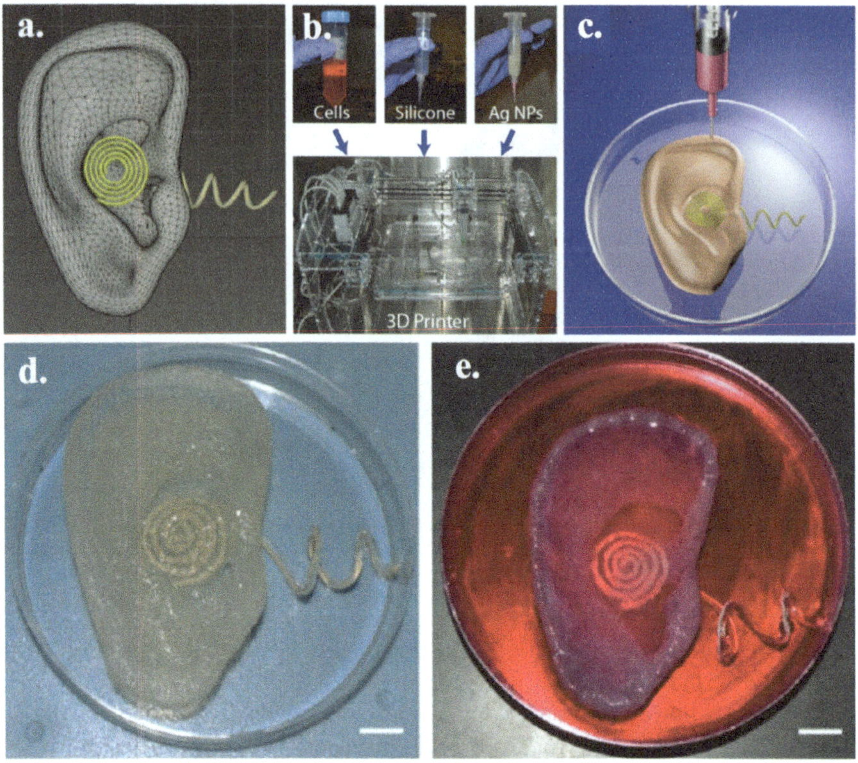

Fig. 9.6 (a) A CAD drawing of the bionic ear. (b) Images of the functional starting materials comprising of; chondrocytes as the biological component; silicone to aid in structural support and; a conducting silver nanoparticle to aid in auditory sensing. (c) An illustration of the 3D printed ear. (d) The 3D printed bionic ear immediately after printing and (e) The 3D printed bionic ear during *in vitro* culture. The scale bars in Figure d and e represent 1 cm. (Adapted with permission from [33])

increasing studies have emerged due to the promising vision of printing *in situ* at the defected or wounded site [34]. *In situ* bioprinting is challenging. There are only limited studies in this niche research field since its first proposal, however, inkjet-based bioprinting has been used for the treatment of burn wounds [35] and a laser-assisted bioprinting technique has been assessed to identify the feasibility of printing nanohydroxyapatite particles onto a wounded mouse model [36]. Ozbolat reported that the bioprinting of skin cells has been considered feasible for the use in a clinical setting for the purpose of repairing major wounds and have shown proof of concept in animal models [37] (Fig. 9.7).

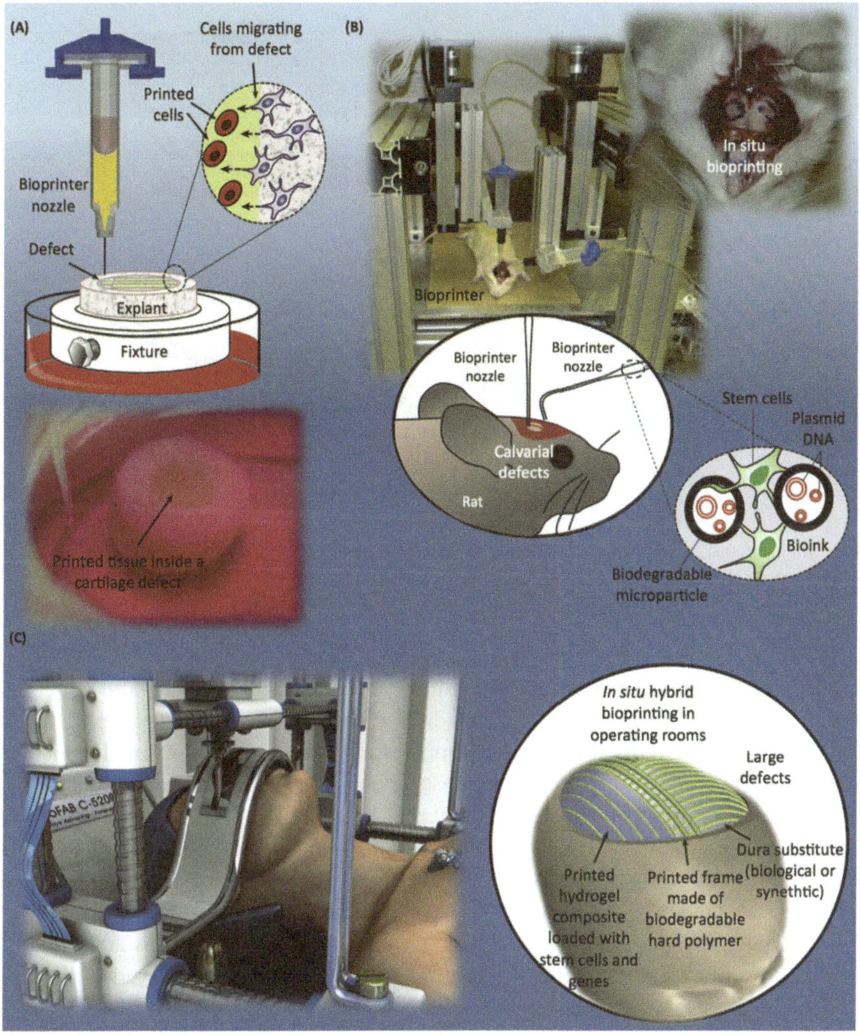

Fig. 9.7 *In situ* model for direct tissue repair. (a) A schematic diagram of *ex vivo* bioprinting process. (b) *In situ* bioprinting of tissue products directly onto animal models. (c) The future vision of transitioning *in situ* bioprinting techniques directly on humans during surgical procedures. (Reprinted with permission from [37])

3.2 The Fabrication of Prosthetics, Implants and Medical Devices

The main product application for the invention of 3D printing was initially prosthetics. Designing and printing personalised implants and prostheses have become the gold-standard method and solution for many patients who require specific

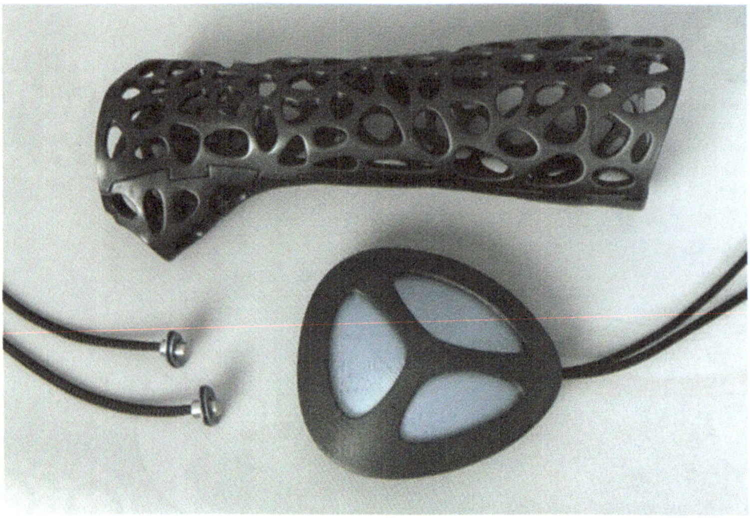

Fig. 9.8 Osteoid™ 3D printed cast prototype that is reported to heal bone injuries faster than a normal cast. (Reprinted with permission from [39, 40])

constructs. Prostheses can be rapidly custom-made using 3D printing both in geometry and material with images from medical scanning (such as radiographs, magnetic resonance imaging (MRI) and computer tomography (CT for example) used to input specifications for production of these devices [38] including dental prostheses alongside part- or full-limb replacements. One example is that of Osteoid™ 3D printed lattice-structured medical casts customised to the wearer (Fig. 9.8). This technique employs ultrasound waves to encourage more rapid bone healing by up to 38% versus standard plaster casts [41, 42]. The cast's ultrasound system (known as LIPUS) pulsates for 20 min per day around the fracture site and has additionally been shown to improve healing of non-union fractures by up to 80% versus conservative methods.

Charitable initiatives are also paving the way for 3D printing applications in regenerative healthcare and provision of surgical prostheses. World's Advanced Saving Project (WASP)® is one such initiative founded in 2012 to develop and supply professional 3D printers in third-world regions for the purposes of encouraging eco-friendly, sustainable manufacturing development, namely, in housing production. Similarly, the Not Impossible™ foundation has also supported and exploited 3D printing as a method of producing and supplying prosthetic limbs and devices to patients in deprived areas.

3.3 Anatomical 3D Models Used for Surgical Training

Sterile surgical instruments have also been successfully printed using 3D printing technology, again setting a precedent for on-demand production of medical equipment. This revelation also comes with numerous economic benefits. Sterile 3D printed surgical retractors, for instance, have been estimated to cost around a tenth of their widely-used stainless steel counterparts [43].

Beyond transplantation and organ bioprinting, however, lie multitudinous opportunities for 3D printing applications in surgery ranging from the modelling of tumours and other abnormal tissue structures *in vitro* to inform surgical approaches. Before the introduction of 3D printing, *in vitro* models were poorly representative of tumour structural complexity and crude approximations of tumour microenvironments. More recently, 3D models have been used to enable more detailed reconstruction of tumour features from cellular proliferation and migration to blood vessel organisation and metastases [44–46]. This also feeds in to the use of 3D printing in transplantation. One case study demonstrated the use of CT scanning in imaging a paediatric patient's airway to subsequently generate a 3D printed tracheal splint [29, 47]. Indeed, this is a useful area of 3D printing in both modelling and in practical utilisation of models to support surgical intervention namely in the generation of splints and guiding templates for resection of bone and other materials [48], as well as suturing devices [49].

In the same vein, the use of 3D printing has extended to the development of targeted tumour therapies, such as in chemotherapy-impregnated mesh devices that can be modelled to specifically fit a given tumour that may otherwise be surgically unachievable and which previously may have spelt the death knell for affected patients. This has already been prototyped in animal models for pancreatic cancer [50] whereby a patient-customised 3D printed bioabsorbable implant is targeted to the tumour site and releases drug at constant therapeutic levels over a period of 4 weeks.

4 FDA Approved Medical Devices

U.S. FDA has granted clearance through the 510(k) process for several 3D printed medical and implantable devices (Table 9.2). The general categories of medical devices that are manufactured by 3D printing include tracheal splints, skill plates, hip prosthetics, spinal cages, dental or bone reconstruction products. In 2017, the FDA has further provided a guidance due to increased investment in the technology and its increased use in medical devices. The purpose of the guidance was to issue technical considerations associated with 3D printing processes [63]. The route to market authorisation of these devices have largely been through the 510(k) pathway where applicants must demonstrate device safety, effectivity and whether it is equivalent or superior to commercial devices currently available. If the FDA approves,

Table 9.2 Examples of FDA approved 3D printed medical or implantable devices*

3D printed medical device	Product	References
Inner ear devices		[51]
Dental crowns	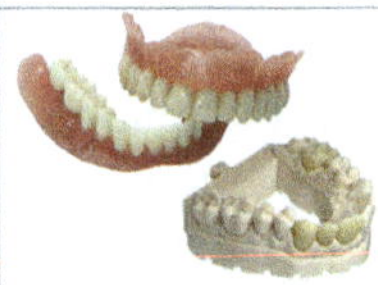	[52]
Bone tether plates	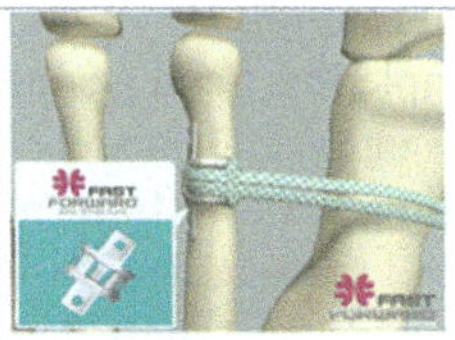	[53, 54]
Skull plates	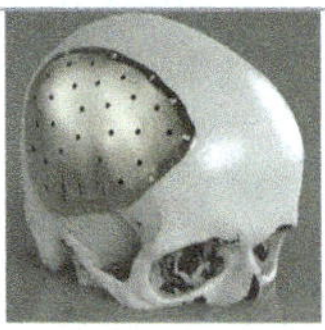	[54, 55]
Spinal plating system		[56, 57]
Facial implants	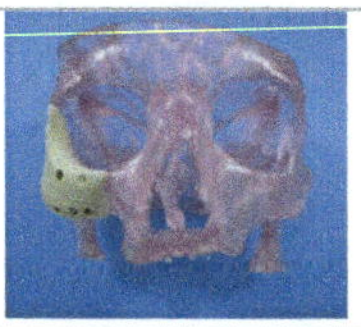	[58, 59]
Surgical instruments	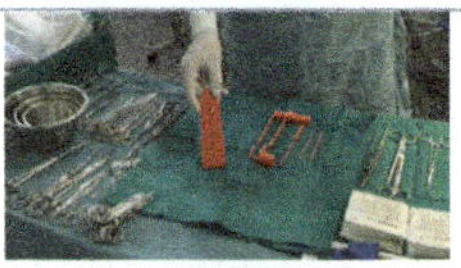	[60, 61]
Invisalign® braces	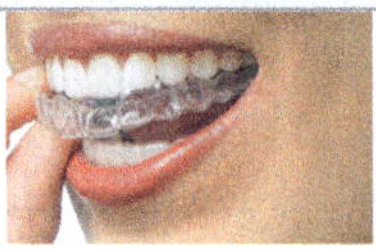	[58, 62]

*Figures reprinted with permission from the references stated the product is said to be "cleared" for commercial use. Although to date of this publication, the FDA has only approved of one 3D printed drug (discussed in Chap. 3), however, the FDA has reportedly cleared at least 85 medical devices made using 3D printing processes. 11 of these cleared 510(k) applications that incorporate 3D printing technologies have been identified. A majority of these 3D printed products either fall in the FDA's Class I or II of medical devices which are low to higher risk devices that require greater regulatory control to provide assurance of device safety and effectiveness [64]. No product, however, has thus far have been approved if classified as a Class III medical device.

Some of these products, like the Invisalign® braces, are 3D printed at a facility and shipped to the prescribing health provider. However, unconventional manufacturing distribution techniques that 3D printing promises have been introduced and implemented in the medical device arena [65]. For example, tracheal splints are able to be printed on-site and on-demand at a health care facility [66], and thus, exemplify the true benefits of rapid prototyping techniques. Companies that fabricate 3D printed medical devices such as FormLabs, EnvisionTEC, Materialise and Renovis Surgical Technologies receive patient specific specifications in the form of a scanned image sent by a physician or a dentist. The digital dentures outlined in Table 9.2 were 3D printed on an EnvisionTEC Vida in the company's FDA- and CE-approved materials including E-Denture (flesh-coloured base) and E-Dent (tooth simulation in a variety of shades). EnvisionTEC has been printing biocompatible materials on its digital light projection printers for more than a decade. Printing the devices in proprietary facilities allow the companies to regulate quality, biocompatibility of materials and sterility.

5 Benefits and Challenges of 3D Printing in Medicine

The benefits of 3D printing also range far beyond its ability to be personalised. Financially, 3D printing offers a competitive alternative to smaller-scale production costs; one example being NASA who utilised 3DP to produce a rocket fuel injector at a third of the cost previously via traditional manufacturing approaches [1]. Other areas of cost-effectiveness extend to areas such as the pharmaceutical industry whereby customisation of specific drugs may enable dose and cost reduction depending on intra-individual profiling for therapeutic dosing [2] and equally in the rapid rate of production of on-demand objects and devices versus multiple-process manufacture in industry or otherwise [31, 67]. This has been exemplified in the generation of 3D printed *in vitro* models considered easier to image, manipulate and process at a higher throughput compared to *in vivo* models [68]. Another important benefit of 3DP and specifically in medicine as well as other fields, is in collaboration and data sharing – a pioneer of this being the National Institutes of Health (NIH), who founded the shared 3D printing data network 3D Print Exchange [69]. Initially used in-house where 3D printers are available throughout the institute's campus for data sharing of software and images for 3DP, this is now an open-access resource enabling all users to share 3D print files for various devices [3, 69].

Although 3D printing offers promising applications and capabilities within the field of medicine, an important obstacle to consider is the feasibility of use in a clinical setting. As with the advent of any other avant-garde technology, and as is especially the case in medicine, regulatory requirements and limitations also apply to the constantly-evolving field of 3D printing, rendering the development of new applications within both an ethically and safety-specific grey area [2, 29]. Whilst printing speeds, processing speeds and resolutions have significantly advanced over the past few years with respect to 3D printing, these parameters have lagged behind the optimal levels often employed for scale-up manufacturing techniques [4]. However, more recently, the FDA have developed a draft guidance to promote the technical considerations specific to devices using additive manufacturing which is promising for the scope of 3D manufacturing [70].

6 Future Advances of 3D Printing in Medicine

The potential for 3D printing as a solution to personalisation as well as on-demand generation of surgical and medical equipment is an exciting and very real one, aside from its other potential in the personalisation and on-demand printing of medicines and medical devices for individual patient use [71–73] be that via handheld 3D printing devices, use of *in situ* printing approaches with precise dimensional specifications, or large-scale 3D printing manufacture. Others have postulated that harnessing stem cells at birth or in early life could also allow for bioprinting of regenerative tissues via the medium of 3D printing [67].

Whilst advances are constantly being made insofar as this notion is concerned and preliminary data is promising, the technology is still primitive for the purposes of printing fully-functional human organs and vascular networks, for instance.

Another aspect is the adoption of 3D printing by the pharmaceutical industry as a viable alternative [71, 74] – many would argue that the purpose of 3D printing is not to replace mass-scale production but to complement it for instance, in the generation of drugs with narrow therapeutic indices or biological products such as antibody-based medicines (see Chap. 3). Whether this is ultimately adopted as such an approach remains to be seen, though the ever-growing research and lay-person utilisation of such devices would suggest in favour.

7 Conclusion

3D printing is an extraordinarily promising and revolutionary technology which has applications across an immensely broad commercial, as well as in a research-based environment. Although its applications are limited by the lag in software and technology development, the opportunities for future uses are innumerable; from bioprinting of human tissues to on-demand generation of prostheses and other medical equipment, not least its applications in other fields with the same pressing need for personalisation. The 3D printing field is ultimately exciting and endless in opportunities with the FDA encouraging the development of science and risk based approaches.

References

1. Schubert C, van Langeveld MC, Donoso LA. Innovations in 3D printing: a 3D overview from optics to organs. Br J Ophthalmol. 2014;98(2):159–61.
2. Ursan ID, Chiu L, Pierce A. Three-dimensional drug printing: a structured review. J Am Pharm Assoc. 2013;53(2):136–44.
3. Gross BC, Erkal JL, Lockwood SY, Chen CP, Spence DM. Evaluation of 3D printing and its potential impact on biotechnology and the chemical sciences. Anal Chem. 2014;86(7):3240–53.
4. Kalaskar DM. 3D printing in medicine. Duxford: Woodhead Publishing; 2017.
5. Williams DF. On the mechanisms of biocompatibility. Biomaterials. 2008;29(20):2941–53.
6. Helsen JA, Breme HJ. Metals as biomaterials. Chichester: Wiley; 1998.
7. Cakor A. Alphaform – Production of hip implant using additive manufacturing: eos; 2014. Available from: https://www.eos.info/press/case_study/additive_manufactured_hip_implant.
8. Developments MD. Top ten: upcoming products to benefit the heart 2011. Available from: http://www.medicaldevice-developments.com/features/featuretop-ten-upcoming-projects-to-benefit-the-heart/.
9. Hench LL, Polak JM. Third-generation biomedical materials. Science. 2002;295(5557):1014–7.
10. Kolan K, Liu Y, Baldridge J, Murphy C, Semon J, Day D, et al. Solvent based 3D printing of biopolymer/bioactive glass composite and hydrogel for tissue engineering applications. Procedia CIRP. 2017;65:38–43.
11. Hench LL, Wilson J. Introduction. An introduction to bioceramics. Singapore: World Scientific Publishing; 2011. p. 1–24.
12. Jungst T, Smolan W, Schacht K, Scheibel T, Groll J. Strategies and molecular design criteria for 3D printable hydrogels. Chem Rev. 2016;116(3):1496–539.
13. Wang J, Goyanes A, Gaisford S, Basit AW. Stereolithographic (SLA) 3D printing of oral modified-release dosage forms. Int J Pharm. 2016;503(1–2):207–12.
14. Martinez PR, Goyanes A, Basit AW, Gaisford S. Fabrication of drug-loaded hydrogels with stereolithographic 3D printing. Int J Pharm. 2017;532(1):313–7.
15. Martinez PR, Goyanes A, Basit AW, Gaisford S. Influence of geometry on the drug release profiles of stereolithographic (SLA) 3D printed tablets. AAPS Pharm Sci Tech. 2018; https://doi.org/10.1208/s12249-018-1075-3.
16. Jones JR. Review of bioactive glass: from Hench to hybrids. Acta Biomater. 2013;9(1):4457–86.
17. Schmidt M, Pohle D, Rechtenwald T. Selective laser sintering of PEEK. CIRP Ann. 2007;56(1):205–8.
18. Miller JS, Stevens KR, Yang MT, Baker BM, Nguyen DH, Cohen DM, et al. Rapid casting of patterned vascular networks for perfusable engineered three-dimensional tissues. Nat Mater. 2012;11(9):768–74.
19. Leukers B, Gulkan H, Irsen SH, Milz S, Tille C, Schieker M, et al. Hydroxyapatite scaffolds for bone tissue engineering made by 3D printing. J Mater Sci Mater Med. 2005;16(12):1121–4.
20. Ventola CL. Medical applications for 3D printing: current and projected uses. Pharm Ther. 2014;39(10):704–11.
21. Pharmaceutials A. Manufactured using 3D printing 2015. Available from: http://www.spritam.com/ - /hcp/zipdose-technology/manufactured-using-3d-printing.
22. Peng W, Datta P, Ayan B, Ozbolat V, Sosnoski D, Ozbolat IT. 3D bioprinting for drug discovery and development in pharmaceutics. Acta Biomater. 2017;57:26–46.

23. Transplantation USGIoODa. Organ Donation Statistics 2016. Available from: https://www.organdonor.gov/statistics-stories/statistics.html.
24. Klein GT, Lu Y, Wang MY. 3D printing and neurosurgery--ready for prime time? World Neurosurg. 2013;80(3–4):233–5.
25. Cui X, Boland T, D'Lima DD, Lotz MK. Thermal inkjet printing in tissue engineering and regenerative medicine. Recent Pat Drug Deliv Formul. 2012;6(2):149–55.
26. Sharma J. Celprogen introduces 3D printed scaffold human heart that may someday be utilized in heart transplants 2016. Available from: http://www.celprogen.com/news-details.php?nid=52.
27. Davis S. Celprogen Inc successfully 3D print functioning pancreas model: TCT Magazine; 2016. Available from: https://www.tctmagazine.com/3d-printing-news/celprogen-successfully-3d-print-functioning-pancreas/.
28. Schmid F. Testing a soft artificial heart Eidgenössische Technische Hochschule Zürich: ETH Zurich; 2017. Available from: https://www.ethz.ch/en/news-and-events/eth-news/news/2017/07/artificial_heart.html.
29. Bartlett S. Printing organs on demand. Lancet Respir Med. 2013;1(9):684.
30. Ozbolat IT, Yu Y. Bioprinting toward organ fabrication: challenges and future trends. IEEE Trans Biomed Eng. 2013;60(3):691–9.
31. Mertz L. Dream it, design it, print it in 3-D: what can 3-D printing do for you? IEEE Pulse. 2013;4(6):15–21.
32. Haumont T, Rahman T, Sample W, MK M, Church C, Henley J, et al. Wilmington robotic exoskeleton: a novel device to maintain arm improvement in muscular disease. J Pediatr Orthop. 2011;31(5):e44–9.
33. Mannoor MS, Jiang Z, James T, Kong YL, Malatesta KA, Soboyejo WO, et al. 3D printed bionic ears. Nano Lett. 2013;13(6):2634–9.
34. Campbell PG, Weiss LE. Tissue engineering with the aid of inkjet printers. Expert Opin Biol Ther. 2007;7(8):1123–7.
35. Skardal A, Mack D, Kapetanovic E, Atala A, Jackson JD, Yoo J, et al. Bioprinted amniotic fluid-derived stem cells accelerate healing of large skin wounds. Stem Cells Transl Med. 2012;1(11):792–802.
36. Virginie K, Fabien G, Isabelle A, Bertrand G, Sylvain M, Joëlle A, et al. In vivo bioprinting for computer- and robotic-assisted medical intervention: preliminary study in mice. Biofabrication. 2010;2(1):014101.
37. Ozbolat IT. Bioprinting scale-up tissue and organ constructs for transplantation. Trends Biotechnol. 2015;33(7):395–400.
38. Rengier F, Mehndiratta A, von Tengg-Kobligk H, Zechmann CM, Unterhinninghofen R, Kauczor HU, Giesel FL. 3D printing based on imaging data: review of medical applications. Int J Comput Assist Radiol Surg. 2010;5(4):335–41.
39. Bilemjian T. Heal bones faster with the 3D printed 'Osteoid' cast: Buro 24/7; 2014. Available from: http://www.buro247.me/lifestyle/technology/heal-bones-faster-3d-cast.html.
40. Osteoid. Osteoid. A better healing experience: Osteoid; 2017. Available from: http://www.osteoid.com/.
41. Watanabe Y, Matsushita T, Bhandari M, Zdero R, Schemitsch EH. Ultrasound for fracture healing: current evidence. J Orthop Trauma. 2010;24(Suppl 1):S56–61.
42. Mundi R, Petis S, Kaloty R, Shetty V, Bhandari M. Low-intensity pulsed ultrasound: fracture healing. Indian J Orthop. 2009;43(2):132–40.
43. Rankin TM, Giovinco NA, Cucher DJ, Watts G, Hurwitz B, Armstrong DG. Three-dimensional printing surgical instruments: are we there yet? J Surg Res. 2014;189(2):193–7.
44. Abbott A. Cell culture: biology's new dimension. Nature. 2003;424(6951):870–2.
45. Griffith LG, Swartz MA. Capturing complex 3D tissue physiology in vitro. Nat Rev Mol Cell Biol. 2006;7(3):211–24.
46. Yamada KM, Cukierman E. Modeling tissue morphogenesis and cancer in 3D. Cell. 2007;130(4):601–10.

47. Zopf DA, Hollister SJ, Nelson ME, Ohye RG, Green GE. Bioresorbable airway splint created with a three-dimensional printer. N Engl J Med. 2013;368(21):2043–5.
48. Song K, Yeom E, Lee SJ. Real-time imaging of pulvinus bending in *Mimosa pudica*. Sci Rep. 2014;4:6466.
49. Tibbits S. 4D printing: multi-material shape change. Archit Des. 2014;84(1):116–21.
50. Yi HG, Choi YJ, Kang KS, Hong JM, Pati RG, Park MN, et al. A 3D-printed local drug delivery patch for pancreatic cancer growth suppression. J Control Release. 2016;238:231–41.
51. tec e. The top choice for 3D printed hearing aids, Inner-ear devices EnvisionTEC; 2017. Available from: https://envisiontec.com/3d-printing-industries/medical/hearing-aid/.
52. tec e. EnvisionTEC dental 3D Printers: accurate, fast, reliable and flexible: EnvisionTEC; 2017. Available from: https://envisiontec.com/3d-printing-industries/medical/dental/.
53. Wheeler A. Medshape INC. Receives FDA clearance for 3D printed Titanium medical device: 3D Printing Industry; 2015. Available from: https://3dprintingindustry.com/news/medshape-inc-receives-fda-clearance-3d-printed-titanium-medical-device-41644/.
54. FDA. 510(k) Premarket Notification: U.S. Food and Drug Administration; 2013. Available from: https://www.accessdata.fda.gov/scripts/cdrh/cfdocs/cfpmn/pmn.cfm?ID=K121818.
55. Revishaw. Surgeons present Renishawplant technology at CMF surgery masterclass: Revishaw Apply Innovation; 2014. Available from: http://www.renishaw.com/en/surgeons-present-renishaw-implant-technology-at-cmf-surgery-masterclass--42387.
56. FDA. 510(k) Premarket Notification: U.S. Food and Drug Administration; 2015. Available from: https://www.accessdata.fda.gov/scripts/cdrh/cfdocs/cfpmn/pmn.cfm?ID=K143126.
57. Medicrea. Lumbosacral anterior plating system: Medicrea; 2017. Available from: https://www.medicrea.com/en/th-lumbar-range/stabolt/.
58. FDA. 510(k) Premarket Notification: U.S. Food and Drug Administration; 2014. Available from: https://www.accessdata.fda.gov/scripts/cdrh/cfdocs/cfpmn/pmn.cfm?ID=K133809.
59. Mendoza HR. FDA approved first 3D printed facial implants: 3DPrint.com; 2014. Available from: https://3dprint.com/12417/3d-print-face-implant-fda/.
60. https://www.invisalign.co.uk/en/Pages/Home.aspx. Invisalign innovations. Pushing what's possible.: Invisalign; 2017. Available from: https://www.invisalign.co.uk/en/what-is-invisalign/Pages/Technology.aspx.
61. FDA. FOIA Response: Align Technology 2011. Available from: https://www.fda.gov/cdrh/510k/K081960.pdf.
62. Málaga HRd. Cirujanos crean instrumentación quirúrgica con impresión 3D: Diario Medico; 2016. Available from: http://www.diariomedico.com/2016/12/05/area-profesional/gestion/cirujanos-crean-instrumentacion-quirurgica-con-impresion-3d.
63. FDA. Technical considerations for additive manufactured devices: Food and Drug Administration; 2017. Available from: https://www.fda.gov/downloads/MedicalDevices/DeviceRegulationandGuidance/GuidanceDocuments/UCM499809.pdf.
64. FDA. Classify your medical device: U.S. Food and Drug Administration; 2014. Available from: https://www.fda.gov/medicaldevices/deviceregulationandguidance/overview/classifyyourdevice/default.htm.
65. Baird LaJ, M. 3D printing: its impact on medical device and healthcare. In: Beek S, Daubert G, Letoureau C, Madagan K, Maiden T, Oni Y, Quinn T, Schryber J, editors. 3D printing of medical devices: when a novel technology meets traditional legal principles. 1st ed. Reed Smith; 2015. p. 1–33.
66. Smith S. 3-D Printing helps save dying baby: CNN; 2013. Available from: http://edition.cnn.com/2013/05/22/health/baby-surgery.
67. Banks J. Adding value in additive manufacturing: researchers in the United Kingdom and Europe look to 3D printing for customization. IEEE Pulse. 2013;4(6):22–6.
68. Burg T, Cass CA, Groff R, Pepper M, Burg KJ. Building off-the-shelf tissue-engineered composites. Philos Trans A Math Phys Eng Sci. 2010;368(1917):1839–62.
69. Health NIo. 3D Print Exchange. Available from: https://3dprint.nih.gov/.

70. Administration FaD. Technical considerations for additive manufactured devices: Draft guidance for Industry and Food and Drug Administration Staff 2016. Available from: https://www.fda.gov/downloads/MedicalDevices/DeviceRegulationandGuidance/GuidanceDocuments/UCM499809.pdf.
71. Trenfield SJ, Awad A, Goyanes A, Gaisford S, Basit AW. 3D printing pharmaceuticals: drug development to front-line care. Trends Pharmacol Sci. 2018;39(5):440–51.
72. Muwaffak Z, Goyanes A, Clark V, Basit AW, Hilton ST, Gaisford S. Patient-specific 3D scanned and 3D printed antimicrobial polycaprolactone wound dressings. Int J Pharm. 527(1–2):161–70.
73. Awad A, Trenfield SJ, Gaisford S, Basit AW. 3D printed medicines: A new branch of digital health. Int J Pharm. 2018;548(1):586–96.
74. Awad A, Trenfield SJ, Goyanes A, Gaisford S, Basit AW. Reshaping drug development using 3D printing. Drug Discov Today. 2018; https://doi.org/10.1016/j.drudis.2018.05.025.

Chapter 10
3D Printing Techniques in the Pharmaceutical Sciences – Intellectual Property Issues

Catherine Jewell and James Stones

Abstract Additive manufacturing, specifically three-dimensional printing (3DP), has been in use in the manufacture of medical devices for some time, with the FDA recently issuing draft guidance for technical considerations in this regard. More recently however, attention has turned to the use of these techniques in modern medicinal manufacture, in particular, orally administrable dosage forms and drug-loaded implants. In addition to the technical challenges faced by inventors in this field, there are a number of legal issues which are likely to develop as 3D printing becomes more widespread, more specifically regarding not only the procurement and enforcement of intellectual property rights, but also potentially arising at the interface of 3D printed pharmaceuticals with competition and consumer protection laws. Intellectual property law can be complex and varies from country to country. This chapter provides an introduction to intellectual property and a discussion of the aspects and issues that are most likely to be relevant to those working in the field of 3D printing of pharmaceuticals.

Keywords 3D printing · Intellectual property · Patents · Trade marks · Designs · Infringement

1 3D Printing Techniques in the Pharmaceutical Sciences & Intellectual Property

This chapter provides an introduction to intellectual property and a discussion of the aspects and issues that are most likely to be relevant to those working in the field of 3D-printing of pharmaceuticals. The aim is to provide a useful summary and a working guide for the lay-person, rather than an intellectual property textbook. For specific advice, it is recommended to consult a qualified patent, trade mark and/or design attorney.

C. Jewell (✉) · J. Stones
Beck Greener, London, UK
e-mail: cjewell@beckgreener.com

A. W. Basit, S. Gaisford (eds.), *3D Printing of Pharmaceuticals*, AAPS Advances in the Pharmaceutical Sciences Series 31,
https://doi.org/10.1007/978-3-319-90755-0_10

1.1 Intellectual Property (IP) Rights

IP Rights are territorial, in other words, governed by national laws which vary from jurisdiction to jurisdiction. Thus, it is generally necessary to consider these on a country-by-country basis. We have focussed here on the law of the UK and Europe, with mention of international systems where these are available and relevant.

The term "IP Rights" is used in this chapter to refer to patents, trade marks, designs and copyrights. However, you may also see this term used elsewhere with a broader interpretation, also including, for example, trade secrets, database rights, confidential or proprietary information, and know-how.

For most IP Rights, it is useful to keep in mind that what is granted to the owner of such a right is the ability to *prevent* third parties from taking certain actions.

In practice, the question of whether you are free to use an invention, design or trade mark is determined by whether or not there are pre-existing rights that could be used to prevent you from doing so. This question generally needs to be considered separately from the question of whether you yourself might be able to obtain an appropriate IP Right, providing you with the ability to prevent third parties from using the invention, design or trade mark.

Similarly, the question of whether you have created a work in which copyright resides is separate from the question of whether your creation of that work infringed a third party's copyright.

1.1.1 Patents

Patents protect inventions: new products and processes which are developments that would not have been obvious to one skilled in the art in the particular field.

The grant of a patent provides the owner with a limited-term monopoly (generally 20 years from the filing date), during which they have the right to prevent third parties from taking certain actions. This is discussed in detail in the section on patent infringement.

The scope of the monopoly is defined by the "claims" of the granted patent.

Broadly speaking, inventions in the UK are defined and claimed as products or processes. However, certain fields of technology tend to use particular claim types, for example "medical use" claims, which have a specific format.

Product claims can include claims which are directed to, for example, formulations, compositions, filaments, intermediate products and final dosage forms. It is also possible to claim an apparatus or system, both of which are technically product claims but which can be viewed as implying additional limitations or features in certain jurisdictions.

Process claims tend to cover processes for production or methods for manufacture. In the UK and Europe, a granted process claim will also provide protection for the direct product of the patented process.

There are also "use" claims and, as mentioned above, medical use claims which are a subset of use claims. It is possible also to claim a "purpose-limited" product, also known as a non-medical use claim, where the use of a known product provides a new technical effect.

Pharmaceutical patent applications often also include more unusual claim types, such as claims to particular dosage regimes, combination claims, and/or claims to methods of drug delivery.

The UK and Europe exclude methods of treatment of the human or animal body from patentability, but other jurisdictions, such as the US, allow these types of claims.

Early 3D-printing patent applications (except for those associated with the development of 3D-printing itself) related to its use in the rapid manufacture of components and prototypes. For example, US 5,204,055, in the name of the Massachusetts Institute of Technology (MIT), relates to layers of powder material bound by a liquid binder selectively deposited to produce a layer of bonded powder material. Repetition of the steps produces a layered bonded component. MIT followed this case with an international application, published as WO 95/11007, which related to "solid free form" methods for making medical devices for controlled release of bioactive agents and implantation. This encompassed techniques such as stereolithography (SLA), selective laser sintering (SLS) and fused deposition modelling (FDM).

However, by the time they filed WO 03/037607, MIT had moved into compression of a 3D-printed dosage form to decrease void space and incorporate a greater amount of active agent (still using liquid binder 3D-printing methods). Cases such as US 2004/0005360 also focussed on the dosage form produced – again, using a liquid binder method where a powder bed was bound into core and shell form by liquid droplets controlled by 3D-printing.

More recent patent applications, for example WO 2014/075185 and WO 2015/143553 by the Orthopaedic Innovation Center, Inc. (CA), continue to focus on implants, in these cases, antimicrobial implants, fabricated using 3D-printing methods.

However, there have now also been a small number of patent applications published which relate to the direct production of complete pharmaceutical dosage forms by 3D-printing methods, for example US 2014/0271862, WO 2016/038356 and WO 2017/010938.

US 2014/0271862 and WO 2017/010938 contain claims to *inter alia* the dosage forms and methods for delivering and producing these. In other words, conventional pharmaceutical-type claims.

In contrast to the more conventional claims of US 2014/0271862 and WO 2017/010938, a recently published international (PCT) patent application by the University of Central Lancashire, WO 2016/038356, illustrates the broader range of claim types and aspects that could usefully be claimed for 3D-printing-related pharmaceutical inventions.

Among other things, this application claims a solid dosage form printing apparatus [1], a method of printing a solid dosage form [2], a solid dosage form obtainable by the method [3], a method of preparing a printing filament [4], the printing filament itself and a filament spool [5], a computer for operating the apparatus [6], and a computer-implemented method of operating the apparatus [7].

The variety of claim types that are potentially available in respect of the different aspects of 3D-printing processes and products provide proprietors with a number of tools for protecting their position. These are discussed below, although it should be noted that these approaches are as yet untested before the courts.

1.1.2 Designs

Designs protect the appearance of products, and in some instances their decoration or ornamentation.

With regard to 3D printed pharmaceuticals, registered and unregistered designs are most likely to be useful with respect to unusually shaped products, provided that the shape is aesthetic, rather than having a technical or functional effect.

UK law provides for registration of designs [8] and also provides for unregistered design rights [9]. EU law allows designs to be registered too (as so-called Registered Community Designs), as well as providing a different form of unregistered right (known as Unregistered Community Designs [10]).

UK unregistered design right only applies to the shape and configuration (how different parts of a design are arranged together) of objects. To enforce unregistered design right you are likely to need proof of when you created a design, for example certified signed and dated design drawings or photos.

UK unregistered design right automatically protects your design for 10 years from the end of the year goods embodying it were first sold, or 15 years from the end of the year the design was created, whichever ends earlier. However, unregistered design right only allows you to object to behaviour resulting from *copying* of your design. If someone else develops the same design independently, then unregistered design right will not allow you to prevent or stop their use.

Registered designs and Registered Community Designs are potentially broader, covering the appearance of the whole or a part of a product resulting from the features of, in particular, the lines, contours, colours, shape, texture and/or materials of the product itself and/or its ornamentation. A "product" is defined as meaning any industrial or handicraft item, including parts of a more complex product, packaging, get-up, graphic symbols and typographic typefaces. Registered designs can protect more than one aspect of a design, for example both the shape and decoration of a product. Both UK and EU design registrations can last for up to 25 years and these are renewable every 5 years.

Unregistered Community Designs protect the same aspects as Registered Community Designs. However, the right arises automatically from the date that the design is first made available to the public within the Community. The term of protection for an Unregistered Community Design is only 3 years from that date.

As in the UK, owners of a Registered Community Design are protected against use of similar designs even when the infringing design has been developed without copying, or even knowledge of the earlier design. In contrast, an Unregistered Community Design grants the right to prevent commercial use of a design only if that design is a copy of the protected one, made in bad faith, i.e. knowing of the existence of the earlier design, as is the case for UK unregistered design right.

The impact on existing Registered Community Designs of the United Kingdom's exit from the European Union is yet to be determined, but it is expected that the UK government and the UK Intellectual Property Office (UKIPO) will ensure that measures are put in place for existing Registered Community Designs to continue to have effect in the UK after the UK leaves the EU, or that equivalent protection will be granted such that owners are not deprived of rights.

1.1.3 Copyright

Copyright protects the form of expression of an idea. In the UK copyright arises upon the following being recorded in a material form:

- original literary, dramatic and musical works
- original non-literary written works, such as software, web content and databases;

and automatically upon creation of:

- sound and music recordings;
- film and television recordings;
- broadcasts; and
- the layout of published editions of written, dramatic and musical works.

You can mark your work with the copyright symbol (©), your name and the year of creation – although you have the same level of protection in the UK regardless of whether or not you have marked your work in this way.

Turning specifically to 3D-printing, copyright is most likely to be relevant with respect to computer aided design (CAD) files and other software, such as software for running CAD programmes or for running 3D printers.

Copyright gives owners the right to prevent people from: copying their work; distributing copies of it (whether free of charge or for sale); renting or lending copies of their work; performing, showing or playing their work in public; making an adaptation of their work; or putting it on the internet.

However, like unregistered design rights, copyright does not provide a full monopoly, but only protects against actions resulting from the copying of your work.

1.1.4 Trade Marks

Under UK law, a "trade mark" means any sign capable of being represented graphically which is capable of distinguishing the goods or services of one undertaking from those of other undertakings. A trade mark may, in particular, consist of words (including personal names), designs, letters, numerals, or the shape of goods or their packaging.

Trade mark law is relevant to all aspects of branding, in particular the name, packaging or shape of a product and/or the name, logos etc. of a company. All of these aspects are likely to be relevant when launching a new company or offering new products or services, hence all are likely to be relevant to 3D-printed products and to companies offering 3D-printed goods and/or services relating to 3D-printing.

A trade mark registration provides the owner with exclusive rights in the trade mark, which are infringed by use of the trade mark in the UK without the owner's consent and within the meaning of Section 10 of the UK Trade Marks Act 1994 (UKTMA 1994).

Similar provisions apply in respect of European Union Trade Mark (EUTM) registrations, obtained through the European Union Intellectual Property Office, which provide a single trade mark registration covering the whole of the European Union. At present this includes the UK. The impact on existing EUTM registrations of the United Kingdom's exit from the European Union is yet to be determined, but it is expected that the UK government and the UK Intellectual Property Office (UKIPO) will ensure that measures are put in place for existing EUTM registrations to continue to have effect in the UK after the UK leaves the EU. Several possible solutions are under discussion, for example a re-registration or extension system, by means of which existing EUTM registrations would be entered on the UK Register and would continue to have effect in the UK, without leaving a gap in coverage.

There are also "international" trade mark registrations which can be obtained through the "Madrid" system. These typically designate multiple jurisdictions, each designation of the resulting registration being treated in the same way as a national trade mark registration in that jurisdiction (assuming each jurisdiction decides to grant protection to the mark in their territory). For example, an International Registration designating the UK would be treated in the UK as having the same effect as a UK registration once the UK Registry grant the designation protection.

Trade marks are registered with respect to specific goods and/or services, set out in the application for registration. Goods and services are divided into "classes". There is an international classification system (the Nice Classification [11]) which is updated on a regular basis. International Class 5 includes mainly pharmaceuticals and other preparations for medical or veterinary purposes, and is likely to be of relevance in relation to those providing 3D-printed pharmaceuticals.

UK and EU trade mark registrations are renewable indefinitely, though they can be removed from the register for various grounds, such as the mark not having been used for 5 years, or the mark becoming generic.

In addition to the legal provisions regarding trade mark registration in the UK it is possible for certain rights to be accrued solely through use of a mark. This is an aspect to be kept in mind when developing a mark and in particular when considering "freedom to use" which is discussed in more detail in a later section.

2 Acquisition of IP Rights

2.1 Patents

Patents, like designs, trade marks and copyright, are territorial rights. Thus, it is generally necessary to obtain separate patents in respect of individual countries. However, there are some international treaties and features that are common to the patent law of most countries that can lighten the administrative and cost burdens in this regard.

The first of these is the Paris Convention and its "priority period" [12]. This is a period of 12 months from the first filing of an application in respect of an invention, during which it is possible for the applicant (or their successor in title) to file a second application in any country recognising priority right. To the extent that the second application relates to the same invention as the first application, this subject-matter will be treated as having an effective filing date which is the same as the filing date of the first application. The UK Intellectual Property office (UKIPO) and European Patent Office (EPO) both recognise and provide for a priority period in line with the Paris Convention.

This is significant as the assessment of whether an application is new and inventive (non-obvious) is carried out based on the state of the art, i.e. what was available to the public at the filing date. Thus, subject-matter in the second application which is directly and unambiguously derivable from the first application will be assessed based on what was available to the public at the effective filing date, i.e. the filing date of the first application.

The priority period also means that the applicant has a year to consider which countries have genuine potential as commercial markets, or could be home to competitors, and in which it therefore may be worth seeking patent protection. Without the priority period, the applicant would have to decide before filing the first application on all of the countries in which to pursue patent protection. The applicant would also have to file applications simultaneously in all of these countries, which can be logistically challenging. Thanks to the priority period, the applicant has a year to do market research and/or work on related improvements to their invention that may be worth including in the second application.

The second treaty to be aware of is the Patent Cooperation Treaty (PCT), which provides a system for filing an "international" (PCT) patent application. The UK and the EPO are both contracting states of the PCT, as are many other countries (the PCT now has 152 contracting member states) [13]. It is permitted to claim priority from a first application when filing a PCT application, and it is also permitted to claim priority from a PCT application when filing a second application, which may also be a PCT application.

In a PCT application, the final choice of countries in which the application is to be pursued is typically made 30 months from the priority date. Hence the applicant can use this system to defer making a final decision on where to pursue patent protection, provided the countries of interest are contracting member states of the PCT.

Applications for UK patents may be filed at the UKIPO. The EPO receives and examines European applications.

International applications under the PCT should be filed at the appropriate receiving office according to the nationality/residence of the applicant, or can be filed directly at the International Bureau of the World Intellectual Property Organization (WIPO) provided that the national security provisions in the applicant's home country allow.

The requirements for patentability of inventions involving 3D-printing are identical to those of any other invention. According to Section 1(1) of the Patents Act 1977 (PA77), a UK patent may be granted only for an invention that is new, involves an inventive step, is capable of industrial application, and which relates to subject-matter in respect of which the grant of a patent is not excluded. There are corresponding requirements under Article 52 of the European Patent Convention (EPC).

An invention is new if it does not form part of the state of the art [14]. The state of the art is taken to comprise all information that has at any time before the priority date of the invention been made available to the public anywhere in the world by any means [15].

However, a prior disclosure is only novelty-destroying if it conveys sufficient information for the invention to be reproduced by the hypothetical "skilled person", i.e. if the disclosure is "enabling" [16].

This raises an interesting point for establishing novelty in emerging technologies such as 3D-printing of pharmaceuticals. As the technology becomes generally more widespread, the enablement requirement for a given disclosure is likely to be reduced as the skilled person becomes more likely to be able to fill in any gaps in the total information content of a disclosure by using their own common general knowledge. Thus, it can be expected to become harder over time to establish novelty of further inventions in this area over a given disclosure.

An invention involves an inventive step if it is not obvious to the skilled person over the state of the art [17]. Thus, one of the key issues in assessing inventive step is to identify the skilled person in question. When 3D-printing was first considered for pharmaceuticals, the skilled person would likely have been a pharmaceutical formulation scientist. However, as 3D-printing becomes more common, the skilled person is more likely to be viewed as a team of people. This could include a formulation scientist, but could also include software engineers and designers who work with 3D-printers routinely as part of their design equipment, and possibly also materials scientists and engineers.

As 3D-printing becomes more widespread, it seems increasingly likely that producing an existing pharmaceutical formulation by means of 3D-printing where this involves only the application of existing 3D-printing processes would be considered obvious, particularly where this does not present any difficulties other than those that can be addressed by "workshop variations" which would be routine for the skilled worker. Thus, as this field of technology advances, establishing inventive step is likely to require overcoming a particular problem or alternatively providing a particular advantage, either through the 3D-printing process itself or in the printed product, in comparison with existing formulations or conventional manufacturing processes.

An invention is capable of industrial application if it can be made or used in any kind of industry, including agriculture [18]. 3D-printing clearly satisfies this requirement.

In the UK, subject-matter in relation to which the grant of a patent is excluded is defined as:

(a) a method of treatment of the human or animal body by surgery or therapy, or
(b) a method of diagnosis practised on the human or animal body [19].

3D-printing therefore is not excluded subject-matter and, in principle, it is clear that 3D-printing-related inventions constitute potentially patentable subject-matter in the UK and Europe.

2.2 Designs

Both UK and Registered Community Designs can claim priority from an earlier application for the same design, although the priority period is only 6 months.

In order validly to register your design in the UK, it must fulfil certain criteria; namely it must be new and have individual character. In addition, it must not be offensive, must not make use of certain protected emblems or flags, and must not be an invention or how a product works (which are aspects of IP covered by patents, not designs). The functionality of a design cannot be protected under design right.

UK applications are filed at the UKIPO by providing details of the design, typically with drawings or photographs. The type of product to which the design relates must also be indicated and this is generally done by indicating the appropriate class of goods under the Locarno classification. Sometimes it is appropriate to indicate in the application which features of the design are considered to be new.

A design must be new for a registration to be valid. Broadly speaking, this means that identical designs must not have been disclosed anywhere in the world before the application is filed. Under UK law there are provisions for a 1 year "grace period" prior to filing, where disclosures by the designer do not destroy the novelty of the design. However, it is usually advisable to keep a design confidential until an application has been filed, because not all countries have this kind of grace period.

Applications for design registrations are not searched or examined by the UKIPO. Only a basic formal examination is carried out. This means that registration and publication of a design can usually be expected within a short period, typically a few weeks after filing.

Although there is no examination of design applications by the UKIPO, if an attempt is made to enforce the design registration, in most cases the validity of that registration is challenged by the other side (either by bringing invalidity proceedings before the UKIPO, or as a counterclaim in the main infringement action in Court). The UKIPO or Court will then consider whether the design is new and possesses individual character (before the Courts will consider whether the registration has been infringed).

Applications for Registered Community Designs are filed at the EUIPO.

Like the UKIPO, the EUIPO does not examine applications for Registered Community Designs other than for formal requirements. However, it is possible for third parties to apply to invalidate a Registered Community Design on grounds including lack of novelty or individual character over designs which have been made available to the public. It is therefore advisable to consider these points when filing an application.

Designs are deemed to lack novelty if their features differ only in immaterial details from a known design. A design is considered to have individual character if the overall impression it produces on the informed user differs from the overall impression made by a known design. This "individual character" assessment may require greater differences than the novelty test.

In considering what designs have been made available to the public in order to apply the novelty and individual character tests, there is an exclusion for disclosures which could not reasonably have become known in the normal course of business to the circles specialised in the sector concerned, operating within the EU. Also, like the UK, EU law allows for a 1 year grace period in respect of disclosures by the designer.

Unregistered Community Designs are also subject to the novelty and individual character requirements.

2.3 *Copyright*

Copyright arises upon creation or recordal in the UK (depending on the type of work concerned), but in other countries it is required that copyright is registered, at least for damages to be awarded. Your work could be protected by copyright in other countries through international agreements, such as the Berne Convention.

In most countries, copyright lasts a minimum of the creator's life plus 50 years for most types of written, dramatic and artistic works, and at least 25 years for photographs. Other types of works can have different copyright periods.

2.4 *Trade Marks*

Applications for UK trade mark registrations are filed at the UKIPO. Assuming that the formal requirements are satisfied, the UKIPO examines each application in respect of "absolute grounds" of registrability. This includes, for example, consideration of whether or not a mark has distinctive character, and whether a mark is descriptive, generic or deceptive, or is contrary to public policy/morality. For EU trade mark registrations, the process is similar, but with applications being filed at the EUIPO.

There are some additional considerations to be borne in mind when applying to register a shape mark (three-dimensional trade mark), rather than a traditional word mark or logo/device mark. These may be of particular relevance for those working in the 3D-printing space.

As indicated above, to be registerable, trade marks must be distinctive. This means that it must be possible for consumers to use the mark as a so-called 'badge of origin' (i.e. they can use the mark to tell that goods and services marketed under the mark come from a particular trade source). The legal test for distinctiveness is the same for shape marks as for other marks. However, both the UKIPO and the EUIPO take the view that the average consumer has a different perception of non-traditional trade marks, such as shape marks. The Offices consider that the average consumer may not as readily accept the appearance of the goods themselves as an indication of trade origin because "average consumers are not in the habit of making assumptions about the origin of the products on the basis of their shape or the shape of their packaging in the absence of any graphic or word element" [20].

The test under current case law [21] is whether the mark enables an average consumer, "without conducting an analytical or comparative examination or paying particular attention", to distinguish the goods concerned from those of other traders. The Offices tend to take the view that only a shape mark which departs significantly from the norms or customs of the sector and thereby fulfils its essential function of indicating origin possesses the distinctive character necessary for registration.

It may be easier to show that a shape has the necessary distinctive character where there is evidence that, contrary to the normal assumption made about consumers' habits, consumers in the relevant sector do in fact rely upon the appearance or the shape of the product or its packaging as a means of identifying the origin of the product.

In addition, for shape marks there are further absolute grounds to be examined, such as whether the shape results from the nature of the goods themselves, whether the shape is necessary to obtain a technical result, and whether the shape gives substantial value to the goods [22].

The purpose behind these requirements is to keep free basic shapes of goods that should be available for use by the public, and shapes which consist exclusively of functional features attributable only to the technical result, i.e. where the shape is one which would be adopted for its functional efficiency. The question of whether the shape gives value to the goods is intended to prevent the securing of a permanent monopoly in the type of shape which should be protected under designs legislation.

Another thing to keep in mind when considering trade mark applications is that in the UK there is a requirement that every applicant must be either currently using the trade mark (or allowing someone else to use it with their consent) or have a bona fide (good faith) intention to use it in relation to the goods or services included in the application. There is a declaration on the application form to this effect, and this must be signed by the applicant or their representative.

Instances where the applicant for a UK trade mark registration has no bona fide intention to use the mark in relation to the goods and services listed in an application are arguably examples of "bad faith" practices. "Bad faith" is a ground for refusal of an application or a finding of invalidity in respect of a registration [23]. This means that it is important to consider carefully the list of goods and services to be specified in a UK trade mark application. There is no equivalent requirement for EU applications.

Once an application has been accepted by the UKIPO or EUIPO, it is published for opposition. During this period, third parties may oppose the application, usually on the basis of earlier rights to an identical or similar mark. This is discussed further in the section on freedom to use.

2.5 Passing Off

In some circumstances, the use of a mark in the course of trade in the UK can lead to the accrual of "goodwill" in the mark, even if the mark is not registered as a trade mark. The common law tort of "passing off" can be used by owners of such goodwill to stop third parties using marks where such use constitutes a misrepresentation which results, or may result, in damage to the goodwill. Typically, this manifests in a business which has been using a trade mark for a long time showing that the public associate the mark with them, and that a third party is using a mark which causes the public to believe that there is a connection, thus causing the first business to lose sales.

Names, packaging, get-up and trade dress are some of the possible so-called indicia in which goodwill may be accrued in the UK.

As a result of their development through UK case law, passing off actions often involve a significant evidential burden. It is necessary to establish at least that there is goodwill, that there has been a misrepresentation, and that this has resulted in, or is likely to result in, damage to that goodwill. All of these factors must be established through evidence.

Passing off has been developed through the jurisprudence of the courts, with the UK and the Republic of Ireland being the only EU countries to recognise such claims. However, some other countries in Europe have provisions relating to unfair competition which offer a similar type of protection.

3 Ownership of IP Rights

3.1 Patents

Under UK law, an inventor owns his or her invention, unless it was made under certain circumstances of employment [24].

Essentially, in the absence of a contractual agreement otherwise, an invention made by an employee belongs to his or her employer if:

(a) it was made in the course of the normal duties of the employee or in the course of duties falling outside his or her normal duties, but specifically assigned to him or her, and the circumstances in either case were such that an invention might reasonably be expected to result from the carrying out of his or her duties; or
(b) the invention was made in the course of the duties of the employee and, at the time of making the invention, because of the nature of his or her duties and the particular responsibilities arising from the nature of his or her duties he or she had a special obligation to further the interests of the employer's undertaking.

Any other inventions are taken as belonging to the employee – again excepting the case where there is a contractual agreement otherwise.

Even where a contract does set out a position in respect of ownership which differs from the statutory position, the relevant terms could be found unenforceable if they do not comply with other parts of UK law.

Specifically, Section 42 UKPA 1977 sets out that any term in a contract relating to inventions by an employee, which diminishes the employee's rights in inventions of any description made by him or her afterwards, or in or under patents/applications in respect of those inventions, shall be unenforceable against the employee to the extent that it diminishes their rights. This applies to contracts between the employee and the employer, or with some other person at the employer's request or in pursuance of the employee's contract of employment.

UK law also provides that where an employee makes an invention which belongs to his or her employer and this invention or the subsequent patent turns out to be of outstanding benefit to the employer, the employee can apply for compensation representing his or her "fair share" of the benefit derived by the employer [25].

Applicants in the UK typically derive their rights in the invention either through being the inventor, through employment of the inventor, by contract, or by assignment (transfer) of rights from the inventor.

Ownership of an invention is an aspect of patent law which can vary greatly depending on the jurisdiction. For example, in the US there are no conditions under which an invention made in the course of employment will automatically be owned by an employer. US applicants rely on assignments of rights to acquire ownership of an invention.

Ownership also has some implications for claiming priority. The right to claim priority belongs exclusively to the applicant or their successor in title. Thus, for example, when filing a PCT application claiming priority from a UK application, the PCT application should be filed in the name of the same applicant(s) as the UK application unless there has been an assignment of rights during the priority year (including the right to claim priority) in which case the PCT application should be filed in the name of the new owner of the invention.

This is a complex area of law and it is advisable to seek professional advice regarding assignments of rights, particularly where these are taking place during the priority period.

3.2 Designs

In the UK, the designer is the first owner of design rights, including where the designer has been commissioned to create the design.

However, the designer is not the first owner of any UK design rights where the design is created by an employee in the course of his or her employment. In this case the employer is the first owner of any design right.

Registered Community Designs and Unregistered Community Design rights follow the same rules, except that there is no express provision that commissioned designers are still the first owner of any design right. Again, the above is the default position, absent any contractual agreement to the contrary. When dealing with design rights under UK or EU law, it is advisable to stipulate ownership in the terms of the design contract.

3.3 Copyright

The first owner of copyright depends upon the type of work created.

For literary, dramatic, musical or artistic works, the author/creator is generally the first owner of any copyright. For films, the principal director and the film producer are generally the joint authors and first owners of copyright. For these works, however, there is an exception where the work was made by employees.

The author and first owner of copyright in a sound recording is usually the producer. For a broadcast, the author and first owner of copyright is usually the broadcaster. The publisher is usually the author and first owner of the copyright in a published edition.

In the event that a literary, dramatic, musical or artistic work, or a film, is made by an employee in the course of his or her employment, the default position is that the employer is the first owner of any copyright in the work. This is the case unless there is an agreement to the contrary.

Although the expression "in the course of employment" is not defined by law, the UK courts often address this point during disputes by considering whether the employee was working under a 'contract of service' (e.g. as an employee) or a 'contract for services' (e.g. as a freelancer or independent contractor).

A person working under a 'contract for services' will often retain copyright in any works they produce, unless there is an agreement to the contrary. Hence an employer should keep careful records of who created a work for them and any contractual agreements which were in force at that time.

In circumstances where you request or commission another person or organisation to create a copyright work for you, the first legal owner of copyright would usually be the person or organisation that creates the work, rather than the commissioner of the work. Again, this is the usual default position, unless there is an agreement otherwise.

However, in the event that, for example, ownership of copyright is not dealt with in the contract to commission the work, UK courts have been willing to find that there is an implied licence allowing the commissioner to use the work for the purpose for which it was commissioned. This may not result in a transfer of ownership – the commissioner may only be found to have a non-exclusive licence. Hence it is important, when commissioning works, to establish who owns copyright through an appropriate contract.

In the event that two or more people have created a single work protected by copyright and the contribution of each author is not distinct from that of the other(s), those people may be considered joint owners.

A good example of a situation where joint ownership might arise would be where a person was commissioned to create a website together with one of the commissioning company's employees. Under these circumstances, the person being commissioned and the company would be likely to be joint first owners of copyright in the website. A third party wanting to copy or use a work of joint ownership would need to obtain agreement from all of the copyright owners.

On the other hand, where individual contributions are distinct or separate, each person would be the author of the part they created. A good example of this would be where the music and lyrics of a song are created by two different people. In this case, where you only wish to use one part, such as the lyrics, you may only need permission from the copyright owner of the lyrics.

Ownership of copyright can be transferred. Hence, it is possible for copyright in material that was produced involving contributions from more than one person to be owned by a single person, as a result of appropriate transfers. Collaborators can agree in advance that copyright in what is to be produced should be owned by a single person or body. It is equally possible for licensing arrangements to be agreed in advance.

There are no specific provisions under UK law for ownership of copyright in works produced by students. This means that the general rules apply, i.e. a teacher/lecturer should not be able to claim joint ownership of a student's work unless they have made a substantial contribution to that work. Nonetheless, it is quite common for universities/colleges to request that students assign their copyright to the establishment when registering for their course. Sometimes the university or college may include a royalty-free licence for any works created. In the absence of any such contract or terms, copyright would likely remain with the creator, i.e. the student.

3.4 Trade Marks

The proprietor of a UK trade mark registration should be listed on the UK trade mark register. Ownership of a trade mark registration can be transferred, in whole or in part, provided this is done in writing and in the appropriate form.

Changes in ownership of a UK registration should be recorded on the UK register. This is important for several reasons, including the fact that the rights conferred by a registered trade mark [26] are exclusive to the proprietor of that trade mark registration. Until an assignment is recorded, it is ineffective against third parties later acquiring a conflicting interest in ignorance of the assignment.

It is also important that the register is up to date so that the UKIPO can send correspondence to the correct place, including, for example, notice that a third party is bringing a revocation action and/or reminders to renew your registration.

3.4.1 Trade Marks and Copyright/Designs

It is possible for elements of a trade mark to be covered by other intellectual property rights. For example, it is quite common for a logo/device mark to be a copyright work in itself.

Since copyright is one of the earlier rights which can prevent use of a mark in the UK [27], it is advisable to arrange for any such copyright to be assigned to the owner of the trade mark application/registration as early as possible in the process of developing the trade mark.

There are certain situations, for example where a device was developed as a commissioned work and the commissioner would want to exclude the designer from use of the device, where an assignment or licence in favour of the trade mark owner may be implied. However, disputes in these kinds of situations are typically much simpler to handle when there is an assignment in writing from the creator of the copyright work to the owner of the trade mark.

A similar situation may arise with regard to registered design rights, or unregistered design rights. Again, these situations are most easily addressed by arranging for assignment of such rights to the trade mark owner, as early as possible in the process of developing a trade mark.

3.5 Passing Off/Goodwill

"Goodwill" is defined as the attractive force that brings in customers to a business [28] and is accrued through use of a sign in relation to particular goods or services.

The owner of goodwill has a property right that can be protected by a passing off action.

4 Patent Infringement Issues

4.1 The National Courts

The possible advent of truly "personalised" medicine gives rise to a number of questions and issues in relation to infringement, both direct infringement and indirect or "contributory" infringement.

It is important to remember that patents provide a "negative" right, i.e. the right to stop unauthorised working of the invention. In essence, UK patent law provides that making, importing, disposing of, offering for disposal, keeping or using a patented product in the UK constitutes infringement, where these acts are committed without the consent of the patent proprietor. Likewise, in relation to a patented process, using the process constitutes infringement, while making, importing, disposing of, offering for disposal, or keeping the direct product of the patented process in the UK also constitute infringing acts. These are "direct" infringements.

"Indirect" or "contributory" infringements essentially consist of supplying or offering to supply in the UK, to a person not entitled to work the invention, and without the consent of the proprietor, any means relating to an essential element of the invention, for putting the invention into effect, when it is known, or is obvious to a reasonable person in the circumstances, that those means are suitable for putting, and are intended to put, the invention into effect in the UK.

Since this discussion relates primarily to pharmaceutical products, it is relevant to note that in the UK, pharmacists currently benefit from the protection of an exemption to infringement. Under s.60(5)(c) UKPA 1977, the extemporaneous preparation in a pharmacy of a medicine for an individual in accordance with a prescription given by a registered medical or dental practitioner, or dealing with a medicine so prepared, are acts which are deemed not to constitute patent infringement.

Would pharmacists' activities remain inside the scope of the existing exemption, even if their "extemporaneous preparation" or "dealing with" a medicine involved 3D-printing, in other words actions equating to manufacture of a pharmaceutical dosage formulation?

In *Generics v Warner-Lambert* [29], there is an indication that if doctors or pharmacists were involved in manufacturing the medicaments in question this could be contributory infringement of a "Swiss-type" claim in the sense of providing a means essential to putting the invention into effect. However, medical use claims are no longer acceptable in the "Swiss-type" format, and the scope of the new medical use claim type, "substance X for treatment of disease Y", has not been tested. Moreover, pending applications relating to 3D-printed dosage forms do not generally appear to include medical use claims or "Swiss-type" claims and these may not be appropriate where the invention relates to the method of manufacture or improved properties of the product itself rather than its use in treating a certain condition.

This begs the question: if a patented product or the direct product of a patented process is only made *in situ* by pharmacists, then who could infringe such a patent? Who can a Patentee sue for unauthorised use of its process, or for performing other acts that would normally infringe, in relation to its product? The exemption for pharmacists appears to mean that they would not be vulnerable to patent infringement actions where their actions relate to preparation of prescriptions.

Nonetheless, 3D-printing processes developed to offer more general benefits such as faster production, reduced wastage or lower costs, or, for example, which allow manufacturers to offer a wider range of dosages without compromising on cost-efficiency, could be attractive to mainstream pharmaceutical manufacturers and generics companies. Such manufacturers would not benefit from the exemption that keeps pharmacists from infringing when performing activities associated with filling prescriptions.

Besides process claims, there are a number of other claim types that could be useful in protecting different aspects of 3D-printing technology.

For example, apparatus claims might catch infringing manufacturers, importers and distributors of hardware – either of the apparatus as a whole or individual elements where these are means essential to working the apparatus and the necessary knowledge or expectation of knowledge is present with regard to intent to work the invention. The same applies to other hardware product claims.

As with the methods of manufacture themselves, claims to the use of 3D-printing in manufacture of a pharmaceutical formulation could be problematic to enforce, with regard to the question of who is the infringer – if this is a pharmacist then they are most likely exempted in the UK, although non-pharmacist manufacturers would not benefit from this exemption.

Manufacturers could also fall foul of "medical use" claims and/or claims to methods of drug delivery, where they are manufacturing an essential element of that method, as well as methods of manufacturing a formulation or its starting materials, such as filaments (where these are essential elements).

Process claims, product claims and intermediate product claims could also encompass as infringers the suppliers of 3D-printing apparatus, or specialised starting materials or intermediates, where these are means essential for putting the invention into effect. Although, in *Nestec v Dualit* [30], where the claimed invention was a system consisting of a coffee machine and a capsule adapted for use in that machine, it was held (with reference to earlier cases [31, 32]) that owners of the relevant coffee machine do not "make" the claimed system when they purchase capsules for use in said machine. Following this approach would suggest that sale of e.g. replacement filaments or spools would be unlikely to infringe 3D-printing apparatus claims.

4.2 The Unified Patent Court

The European patent system is expected shortly to undergo its biggest change in 40 years, with the advent of the unitary patent package [33]. This will introduce a new option for patent proprietors, who will be able to request unitary effect upon the grant of a European patent application.

Requesting unitary effect will provide a single European unitary patent covering all of the member states which have signed up to the system at that time. Alternatively, proprietors may continue to use the existing validation system to obtain patents in some or all of these member states, resulting in a bundle of independent national patents that are maintained, challenged and enforced individually.

The unitary patent package is made up of two parts: the unitary patent, and the Unified Patent Court (UPC). Once the system comes into effect, the UPC will have exclusive jurisdiction with regard to the litigation of unitary patents and, for a transitional period, will share jurisdiction over non-unitary European patents in the relevant member states, where these have not been opted out of the UPC. After the transitional period, the UPC will have exclusive jurisdiction over all European patents in respect of the relevant member states, unless these have been opted out of the UPC before the end of the transitional period (and the opt-out has not been withdrawn).

The UPC is not yet open, and the Court's approach to infringement is therefore as yet untested. Nonetheless, there are a few general points worth mentioning.

First, the UPC will be a pan-European court and therefore will have access to pan-European remedies. The benefits for patent proprietors include, for example, the possibility of obtaining a pan-European injunction. On the other hand, the corollary to this is that a unitary patent would be vulnerable to revocation through a single action in the UPC.

Secondly, the Unitary Patent Regulation [34] states that the acts against which a unitary patent provides protection (i.e. infringing acts) and the applicable limitations (defences to infringement) shall correspond to those defined by the law of the participating member state whose national law is applicable to the unitary patent as an object of property, as defined by Article 7 of the Unitary Patent Regulation.

Article 7 states that a unitary patent as an object of property shall be treated as a national patent of the participating member state in which that patent has unitary effect and in which, according to the European Patent Register: (a) the applicant had its residence or principal place of business on the date of filing of the application for the European patent; or (b) where point (a) does not apply, the applicant had a place of business on the date of filing of the application for the European patent. For co-owned patents, point (a) is applied to all owners in turn, in the order in which they are listed in the register. If this does not provide an answer then point (b) is applied in the same order. Should this also not provide an answer then the patent is treated as an object of property according to the laws of Germany.

What this appears to mean is that the UPC will apply an approach to infringement and defences which is consistent with the national law of the relevant member state. All of the participating member states have been required to implement in national law provisions which are equivalent to those laid out explicitly in the UPC Agreement [35], however, these various national implementations may not be identical.

For example, statutory defences to infringement are not identical across all participating member states. In particular, the scope of the UK's regulatory review defence is rather broader than that available in other EU countries such as France and Germany.

The UK has broadened the experimental use exemption, setting out that "anything done in or for the purposes of a medicinal product assessment which would otherwise constitute an infringement of a patent for an invention is to be regarded as done for experimental purposes relating to the subject-matter of the invention" [36]. A "medicinal product" is defined in this context as "a medicinal product for human use or a veterinary medicinal product" [37].

There does not seem to be any restriction placed on the types of medicinal products covered, and although there is no explicit reference to innovative products, it is generally accepted that this provision exempts from infringement trials relating to innovative drugs.

A "medicinal product assessment" is defined as "any testing, course of testing or other activity undertaken with a view to providing data" for purposes including obtaining marketing authorisation [38]. It remains to be seen precisely which acts are exempt under this provision. For example, it is not clear whether the supply of patent-protected products to a company which intends to use the product in tests necessary to obtain marketing authorisation is exempt from patent infringement.

There is a risk that the law on infringement to be applied to Unitary Patents by the UPC will depend on the test in Article 7 of the Regulation. In light of this possibility, the UPC may in practice ignore the requirement to apply the appropriate national law and may apply the UPC Agreement directly. Nonetheless, with regard to 3D-printing and related products, the broader scope of this defence in the UK could be highly relevant with respect to both unitary and non-unitary European patents.

5 Supplementary Protection Certificates

The duration of certain rights associated with a patent may be extended in some cases by means of a Supplementary Protection Certification (SPC). An SPC is typically available for regulated, biologically active agents, such as human or veterinary medicaments and plant protection products, like insecticides and herbicides.

The justification for supplementary protection certificates is that these compensate patent proprietors for the lengthy processes for obtaining regulatory approval of these types of products. Obtaining authorisation to put these products on the market can take several years.

A supplementary protection certificate comes into force only after the corresponding patent expires and usually has a maximum lifetime of 5 years. This can be extended to five and a half years for particular paediatric medicines.

There are no special considerations for SPCs in respect of 3D-printed products. SPCs generally extend the monopoly period for a "product" – an active ingredient or a combination of active ingredients – that is protected by a patent. In some cases, there is no difficulty about the definition of the "product" or whether it is protected by the patent upon which the SPC application was based. However, in other cases, especially concerning medicinal products containing multiple active ingredients, the definition of the "product" is not clear cut, and it is not immediately apparent whether a patent protects that "product". It is generally recommended to seek professional advice in this regard.

6 Trade Marks – Freedom to Operate & Infringement

Trade marks are important tools for businesses to establish an appealing brand for their products. However, when developing a trade mark, attention needs to be given to those names already being used in the marketplace or registered as trade marks, in particular, to avoid similarities that may lead to a likelihood of confusion in the mind of the consumer between the product and identical or similar products from different sources.

Viewed simply, trade mark issues can be divided into those which relate to registration of a mark and those which relate to use of a mark (your own use, and/or that of third parties). There is some overlap between the factors relevant for consideration in respect of registration and use, but it can be helpful to look at them separately, as each has different implications in practice.

Issues relating to registration pertain to whether you can obtain a trade mark registration for your mark(s), i.e. marks which you have developed/created for your own use in commerce. Absolute grounds under which the relevant registry may refuse an application have already been mentioned, but there are also relative grounds – is there an earlier right in existence which entitles its proprietor to prevent the use of your mark? The UKIPO opposition procedure provides a chance for third party rights owners to try and block the registration of your mark, on the basis of, inter alia, their earlier rights. The factors which determine whether the owner of an earlier right can block the registration of your mark are set out in Section 5 UKTMA 1994, and include conflicting marks as defined in Sections 10 (1)–(3) UK TMA, as well as rights under passing off and copyright/design rights. There are equivalent provisions as regards EU applications.

Issues relating to use can be thought of in terms of issues relating to your own use and those relating to use by third parties. Issues relating to your own use naturally tie in with the risk of opposition to your trade mark application(s), and also pertain to possible risks that your use of a mark might infringe a third party's rights. In issues relating to third party use this situation is typically reversed, and you may be concerned about a possible infringement of your rights, by the third party's use of a mark.

6.1 *Freedom to Use*

Before commencing use of a mark it is advisable to conduct "clearance searches". These are typically searches of the official trade mark registers in the relevant countries, which identify trade mark applications and registrations in respect of potentially conflicting marks specifying goods/services in the relevant classes. Searches are usually conducted with respect to particular classes – not only the class(es) in which your goods or services of interest are listed, but also classes which contain potentially similar goods and/or services.

Conducting searches provides for the evaluation of the possible risks of using a mark, based on what is already on the register – for example, there may be an identical mark already registered in respect of the identical goods or services, such that use of your proposed mark would constitute an immediate risk of infringement, and which would represent a significant risk of opposition if you were to apply to register your proposed mark.

It is more difficult to evaluate the potential risks of a passing off action by means of clearance searches, since passing off is a common law tort and does not require a trade mark registration. However, it is sometimes possible to conduct "common law searches" which may identify conflicting unregistered marks that are in use, thus aiding in evaluation of the risks associated with using your proposed mark.

6.2 *Infringement*

A UK trade mark registration provides the owner with exclusive rights in the trade mark, which are infringed by use of the trade mark in the UK without the owner's consent and within the meaning of Section 10 UKTMA 1994, set out below. Article 9 of the EU Trade Mark Regulation [39] provides similar provisions as regards EU trade mark registrations.

The UK provisions [40] are as follows:

1. A person infringes a registered trade mark if he uses in the course of trade a sign which is identical with the trade mark in relation to goods or services which are identical with those for which it is registered.
2. A person infringes a registered trade mark if he uses in the course of trade a sign where because:

   (a) the sign is identical with the trade mark and is used in relation to goods or services similar to those for which the trade mark is registered, or
   (b) the sign is similar to the trade mark and is used in relation to goods or ser vices identical with or similar to those for which the trade mark is registered, there exists a likelihood of confusion on the part of the public, which includes the likelihood of association with the trade mark.

3. A person infringes a registered trade mark if he uses in the course of trade, in relation to goods or services, a sign which is identical with or similar to the trade mark, where the trade mark has a reputation in the United Kingdom and the use of the sign, being without due cause, takes unfair advantage of, or is detrimental to, the distinctive character or the repute of the trade mark.
4. For the purposes of this section a person uses a sign if, in particular, he—

   (a) affixes it to goods or the packaging thereof;
   (b) offers or exposes goods for sale, puts them on the market or stocks them for those purposes under the sign, or offers or supplies services under the sign;
   (c) imports or exports goods under the sign; or

(d) uses the sign on business papers or in advertising.

5. A person who applies a registered trade mark to material intended to be used for labelling or packaging goods, as a business paper, or for advertising goods or services, shall be treated as a party to any use of the material which infringes the registered trade mark if when he applied the mark he knew or had reason to believe that the application of the mark was not duly authorised by the proprietor or a licensee.
6. Nothing in the preceding provisions of this section shall be construed as preventing the use of a registered trade mark by any person for the purpose of identifying goods or services as those of the proprietor or a licensee.

   But any such use otherwise than in accordance with honest practices in industrial or commercial matters shall be treated as infringing the registered trade mark if the use without due cause takes unfair advantage of, or is detrimental to, the distinctive character or repute of the trade mark.

6.2.1 Likelihood of Confusion

Where there is use of a mark which is identical to a registered trade mark, in relation to identical goods or services to those listed in the registration, this is sufficient for there to be infringement of the trade mark registration. In these circumstances, it is not necessary for there to be a likelihood of confusion on the part of the relevant public for the trade mark registration to be infringed.

However, where the marks are similar and the goods/services are identical, or where the marks are identical and the goods/services are similar, or where the marks are similar and goods/services are similar, in order for the trade mark registration to be infringed under s.10(2) there must also be a likelihood of confusion.

Assessment of whether there is a likelihood of confusion between a candidate mark and an earlier trade mark is made from the perspective of the "relevant consumer". Traditionally, the relevant consumer, in relation to prescription-only pharmaceutical products, has been held to be a healthcare professional, such as a doctor, pharmacist or nurse. This was significant, to the extent that they were perceived to display a greater capacity to distinguish between pharmaceutical marks than the end consumer.

In practical terms this meant that a lesser degree of difference was required to distinguish pharmaceutical marks than would otherwise be the case, although the enhanced capacity of these healthcare professionals to distinguish between marks had to be offset against the enhanced risks to public health due to potential medication errors, if confusion did occur.

However, the Court of Justice of the EU has held in certain cases that the relevant public, for the purposes of assessing a risk of confusion between pharmaceutical marks, includes both the end user and the healthcare professional [41]. Although the Court acknowledged that healthcare professionals were likely to influence or even determine the choice of product made by end users, this had to be balanced against the ability of end users to make those professionals take into account their perception of the marks at issue and, in particular, their requirements or preferences.

In another case, the General Court held that the end user would display an "above average" degree of attention when choosing pharmaceutical products, given the risks associated with a medication error [42].

Nonetheless, in circumstances where end consumers are not likely to display as great a degree of attention and appreciation of the differences between pharmaceutical marks as would healthcare professionals, these cases indicate a lowering of the threshold for confusion between pharmaceutical marks. Thus marks for pharmaceuticals are likely now to need a higher level of distinctiveness than perhaps was the case in the past to avoid conflict.

6.2.2 Reputation

There is also provision in the UK for infringement if a mark is used which is identical or similar to a registered trade mark, in relation to any goods or services, provided the registered mark has a reputation in the UK, and use of the sign is (i) without due cause, and (ii) takes unfair advantage of, or is detrimental to, the distinctive character or the repute of the trade mark. Reputation typically must be established by means of appropriate evidence and the evidential burden here is relatively high.

6.2.3 Conclusion

In most commercial areas, trade mark owners are relatively free to select the marks they wish to register and/or use in marketing their goods or services. Generally speaking, provided there is no risk of confusion on the part of the relevant public with an earlier mark, and no identical or similar earlier mark with a reputation, a trade mark usually will be free for use and registration.

There are some additional considerations which are particular to the selection of trade marks for pharmaceutical products, however, and these are discussed in more detail below.

7 Pharmaceutical Trade Marks

In the EU, it is necessary to obtain a marketing authorisation before a medicinal product can be sold, and part of this process involves the assessment of the proposed trade name of the medicinal product. The trade name of the product will form part of any marketing authorisation which is eventually granted. The authorities considering whether to grant a marketing authorisation will evaluate whether the brand, i.e. the chosen name, poses a risk to public health. This is not determined with regard to the trade mark infringement factors discussed above, but by an assessment of the potential for medication errors that could result in serious health risks.

The regulation of pharmaceutical products for the protection of public health has a significant impact on the process of selecting a trade mark. In cases where similar trade marks are used for different medical indications, there is a public health conflict. However, in cases where similar trade marks are used for the same medical indication, there is a trade mark conflict. These issues must be balanced in selection of the final mark, increasing the complexity of the process.

The type of marketing authorisation sought determines the detailed considerations for an acceptable trade mark. There are several types of marketing authorisation available in the EU, including the centralised procedure through the European Medicines Agency (EMA), which is mandatory for certain types of medicines. It is also possible to apply for marketing authorisation through national member state procedures, where the product is not on the EMA's mandatory list. For example, in the UK, the Medicines and Healthcare Products Regulatory Agency sets and regulates the procedure for obtaining marketing authorisation. There is also a decentralised procedure whereby companies can apply for simultaneous marketing authorisations in any number of EU member states, provided the product has not already been authorised in a relevant member state and is not on the EMA's mandatory list. The mutual recognition procedure allows authorisation in a first EU member state to be used to obtain marketing authorisations from other EU member states which recognise the validity of the original marketing authorisation.

The relevant regulatory bodies each have their own guidelines relating to acceptable pharmaceutical trade marks. The EMA, for example, has the "Guideline on the Acceptability of Names for Human Medicinal Products Processed through the Centralised Procedure" which is applied by the Name Review Group of the EMA's Committee for Human Medicinal Products.

According to the EMA's guidelines, "A Community marketing authorisation is valid throughout the EU and the invented name of the medicinal product is an integral part of the authorisation". A crucial point in these guidelines is the requirement for a single trade mark throughout the EU. Although a second name can, in theory, be permitted by the European Commission if the first name proves to be unavailable in a member state, as yet there are no instances in practice where a second name has been permitted under the centralised procedure.

Article 1(20) of Directive 2001/83/EC provides that the name of a medicinal product "may be either an invented name not liable to confusion with the common name, or a common name or scientific name accompanied by a trade mark or the name of the marketing authorisation holder."

7.1 Invented Names

The EMA guidelines explicitly state that only considerations relating to public health or safety will determine whether an invented trade mark can be used for the medicinal product in question – even where the applicant has registered their brand as a trade mark throughout the EU.

The EMA guidelines set out general criteria for reviewing the acceptability of proposed pharmaceutical trade marks, and in summary these should not:

- be liable to cause confusion in print, handwriting or speech with the invented name of another medicinal product (taking into account factors such as the indications, patient population, pharmaceutical form, route of administration, strength, setting for prescription, dispensing and use, and the potential for harm in the event of confusion);
- convey misleading therapeutic and/or pharmaceutical connotations;
- be misleading with respect to the composition of the product;
- convey any promotional message with respect to the characteristics or composition of the product; or
- be offensive or have an inappropriate connotation in any of the official EU languages.

The use of qualifiers/abbreviations by letters as part of an invented name is viewed as acceptable in principle. The use of numbers may also be acceptable in certain cases. There are specific considerations to be applied when assessing the acceptability of qualifiers/abbreviations by letters/numbers, namely:

- whether the qualifier/abbreviation provides further information on the characteristics of the medicinal product or provides for a differentiation, which could help healthcare professionals and/or patients to prescribe/select the appropriate medicinal product; and
- the balance between the potential risk to public health in case of medication error potentially related to the qualifier/abbreviation versus the potential risk resulting from more complex names, adversely affecting in turn memorability, pronunciation and/or prescription.

Qualifiers consisting of a single letter or number are not recommended, as they may be confused with the strength and/or dosage of the medicinal product. Qualifiers/abbreviations which require translation to be understood in the respective EU member state are not acceptable, on the grounds that they are incompatible with the single name rule.

Further additional criteria also apply to combination products and specific product types (e.g. vaccines, orphan drugs (pharmaceutical products which have been developed specifically to treat a rare medical condition) and non-prescription products).

7.2 Common/Generic Names

Rather than using an invented name, it is possible to use the common or scientific name in combination with a trade mark or the name of the marketing authorisation holder.

The common, or generic, name will most often be the International Non-Proprietary Name (INN) for the product. INNs facilitate the identification of pharmaceutical substances or active pharmaceutical ingredients. Each INN is a unique name that is globally recognized and is public property. The World Health Organisation (WHO) administers the system for the allocation of INNs.

7.3 Confusion between an Invented Name and an INN

There are a number of factors that determine whether an invented pharmaceutical brand name will be found confusingly similar to an INN. The EMA's Guidelines recommend that invented names should not be derived from INNs and INN stems should not be used in invented names.

In the event that the EMA considers that an invented name is similar to an existing INN, it will consider the following criteria in determining whether the invented name should be allowed:

- closeness in speech or writing with its own or a different INN;
- similarity in medical setting, general use (indication) of concerned medicinal products;
- similarity in classification for supply of the concerned medicinal products e.g. restricted to hospital setting or specialists;
- route of administration and, where possible, the pharmaceutical forms.

In the event that an invented pharmaceutical name contains an existing INN stem, the Name Review Group will consider the following criteria in determining whether the invented name should be allowed:

- similarity in therapeutic class between the INN stem and the medicinal product;
- location of the INN stem within the proposed invented name, with reference to the WHO INN stem location recommendations;
- similarity in medical setting, general use (indication) of concerned medical products;
- similarity in classification for supply of the concerned medicinal products e.g. restricted to hospital, specialists;
- route of administration and, where possible, the concerned pharmaceutical forms.

7.4 Summary

Regardless of the requirement for a single brand name under the EMA's centralised procedure for obtaining a marketing authorisation, most pharmaceutical companies will wish to secure a global brand in order to maximise recognition of the product.

The challenges and special considerations outlined above in this regard mean that it is advisable to begin the process of creating a trade mark as early as possible, once a product is in clinical trials. In view of the complexity of the regulatory framework governing the use of trade marks for pharmaceutical products and the difficulty of securing a single EU mark, thorough and prompt clearance searches are also a necessary part of this process.

8 Exhaustion of Rights and Parallel Imports

Each member state in the EU decides the prices of the drugs sold in its territory, based on its own healthcare policies. This results in significant differences in the price of pharmaceuticals across the EU, which in turn creates business opportunities for parallel importers who can buy the drugs in countries where they are cheaper and import them into the more expensive countries.

Packaging and labelling of pharmaceuticals is tightly regulated, both at the EU level and by individual member states. A product acquired in one member state therefore often needs to be repackaged before it can be placed on the market in another member state. However, such repackaging can interfere with the trade mark rights of the original supplier of the product.

There is frequently a conflict between the protection of trade marks (and other IP) and the principle of free movement of goods within the EU. Article 34 of the Treaty on the Functioning of the European Union provides for the free movement of goods between EU member states, while Article 36 of this treaty says that prohibitions or restrictions on imports can be justified on the grounds of protecting industrial or commercial property (including trade marks) provided that they do not constitute a means of arbitrary discrimination or a disguised restriction on interstate trade.

The question of what is a disguised restriction vs. what is justified trade mark enforcement is examined under current case law [43] by considering two main questions. Were the goods that the parallel importer wished to import placed on the market by the trade mark owner, or with its consent, or in such a way that the trade mark owner had the opportunity to control their quality? And if the answer to this question is no, was the party that placed the goods on the market under a trade mark also the party in effective control of the trade mark being enforced? If the answer to this question is also no, then it would be expected that enforcement of the trade mark would be considered justified.

If the answer to either of these questions is yes, then the situation may be different, and the proprietor's rights may have been exhausted by the first marketing of the goods within the EU, and hence the trade mark owner can no longer object to the further commercialisation of those goods by others.

Exhaustion of rights does not apply where there are legitimate reasons for the proprietor to oppose further commercialisation of the goods, especially where the condition of the goods is changed or impaired after they have been put on the market [44].

This provision has been interpreted by the CJEU in the *Bristol-Myers-Squibb* case [45] to mean that the trade mark owner cannot legitimately oppose further marketing of a repackaged pharmaceutical if:

- repackaging is necessary to market the product in the country of importation;
- it does not affect the original condition of the product inside the packaging;
- the new packaging clearly states who repackaged the product and the name of the manufacturer;
- the presentation of the repackaged product is not liable to damage the reputation of the trade mark or of its owner; and
- the importer gives notice to the trade mark owner before the repackaged product is put on sale, and, on demand, supplies the owner with a specimen of the repackaged product.

These are known as the five BMS conditions, and these apply both to re-boxing and to over-stickering. If any of these conditions is not satisfied then the trade mark owner can oppose further marketing of the repackaged product.

It is not yet known how the existing law on repackaging might be applied in relation to 3D-printed pharmaceuticals: where a 3D-printed pharmaceutical is licensed for production and sale in one member state, will it be allowable for parallel importers to sell it in another if the proprietor has consented to putting the product on the market in the first member state? If so, it is reasonable to expect that any subsequent repackaging would need to satisfy the same BMS conditions as traditional pharmaceutical products.

It is also worth considering whether the sale of 3D-printing apparatus in an EU state and licence to manufacture would equate to consent to put the pharmaceutical products on the market in that state and therefore exhaust the proprietor's right to stop import of that product into any other EU state. It is well-established by case law that in this context, consent must be expressly given. However, licence to manufacture would appear redundant without a licence to sell the product.

At present, owners of IP rights such as patents and trade marks can use customs provisions as part of their enforcement strategy, arranging for the seizure of suspected infringing imports. However, the mobility of 3D-printing systems may mean unlicensed third parties are more likely to produce infringing products locally, rather than exporting/importing, which would reduce the usefulness of the current customs provisions as part of an overall enforcement strategy.

9 Conclusion

In addition to the technical challenges faced by inventors in this field, there are a number of legal issues which are likely to develop as 3D printing becomes more widespread, in particular regarding not only the procurement and enforcement of intellectual property rights, but also potentially arising at the interface of 3D printed pharmaceuticals with competition and consumer protection laws.

Intellectual property law can be complex, and varies from country to country, so it is recommended to seek professional assistance from a qualified patent, trade mark and/or design attorney, or an IP Solicitor, who will be able to provide detailed advice in relation to your particular situation.

References

1. WO 2016/038356, claims 1–19.
2. WO 2016/038356, claims 20–24.
3. WO 2016/038356, claim 25.
4. WO 2016/038356, claims 26–27.
5. WO 2016/038356, claims 29–31.
6. WO 2016/038356, claim 32.
7. WO 2106/038356, claim 33.
8. Registered Designs Act 1949 (RDA 1949).
9. Copyright, Designs and Patents Act 1988 (CDPA 1988).
10. Community Design Regulation: COUNCIL REGULATION (EC) No 6/2002 of 12 December 2001 on Community designs.
11. http://www.wipo.int/classifications/nice/en/.
12. Article 4, Paris Convention for the Protection of Industrial Property of March 20, 1883, as revised at Brussels on December 14, 1900, at Washington on June 2, 1911, at The Hague on November 6, 1925, at London on June 2, 1934, at Lisbon on October 31, 1958, and at Stockholm on July 14, 1967, and as amended on September 28, 1979.
13. Full list available here: http://www.wipo.int/pct/en/pct_contracting_states.html.
14. Section 2(1) PA77; Article 54(1) EPC.
15. Section 2(2) PA77; Article 54(2) EPC.
16. Merrell Dow v Norton [1996] RPC 76 HL.
17. Section 3(1) PA1977; Article 56 EPC.
18. Section 4(1) PA77; Article 57 EPC.
19. Section 4A(1) PA77; Article 53 EPC.
20. Procter & Gamble vs. OHIM, Joined Cases C-468/01 P to C-472/01 P.
21. Henkel C-218/01.
22. Section 3(2) UKTMA 1994.
23. Section 3(6) and Section 47(4) UKTMA 1994.
24. Section 39–43 UK Patents Act 1977.
25. Section 41 UKPA 1977.
26. Section 9 UKTMA 1994.
27. Section 5 (4) UKTMA 1994.
28. IRC v Muller & Co's Margarine Ltd [1901] AC 217.
29. Generics (UK) Ltd (t/a Mylan) v Warner-Lambert Company LLC [2015] EWHC 2548 (Pat).
30. Nestec SA & Ors v Dualit Ltd & Ors [2013] EWHC 923 (Pat).
31. Schütz (UK) Ltd v Werit UK Ltd [2013] UKSC 16.
32. United Wire Ltd v Screen Repair Services (Scotland) Ltd ([2001] RPC 24).
33. Subject to ratification of the Unified Patent Court Agreement by the necessary parties and subject to the outcome of a pending Court challenge in Germany.
34. REGULATION (EU) No 1257/2012 OF THE EUROPEAN PARLIAMENT AND OF THE COUNCIL of 17 December 2012 implementing enhanced cooperation in the area of the creation of unitary patent protection.
35. Articles 25 to 29 of the UPC Agreement.

36. Section 60(6D) UK Patents Act 1977, as amended by the Legislative Reform (Patents) Order 2014.
37. Section 60(6F) UK Patents Act 1977, as amended by the Legislative Reform (Patents) Order 2014.
38. Section 60(6E) UK Patents Act 1977, as amended by the Legislative Reform (Patents) Order 2014.
39. COUNCIL REGULATION (EC) No 207/2009 of 26 February 2009 amended by Regulation (EU) 2015/2424 of the European Parliament and of the Council of 16 December 2015.
40. Section 10 UKTMA 1994.
41. Case C-412/05 Aventis Pharma v OHIM (26 April 2007).
42. Case T-95/07 Aventis Pharma v Nycomed GmbH (21 October 2008).
43. Flynn Pharma Ltd. Vs. DrugsRUs and Another, Court of Appeal [2017] EWCA Civ 226.
44. Article 7, Directive 2008/95/EC.
45. Joined Cases C-427/93, C-429/93 and C-436/93, Bristol-Myers-Squibb [1996] ECR I-3457, para 79.

Chapter 11
Regulatory Perspectives on 3D Printing in Pharmaceuticals

Akm Khairuzzaman

Abstract The genomic revolution and the age of personalized medicine has exponentially increased in interest and understanding since its introduction in the last decade. Embracing innovation, the United States Food and Drug Administration (FDA) approved the first three-dimensional (3D) printed drug product and thus, has set the benchmark for an industrial revolution in pharmaceutical manufacture. Regulatory standards, however, require each product of each batch to be within specifications with respect to purity, impurities and quality. The regulatory environment of individual countries, however, differ in terms of manufacturing, distribution and compounding of drug products and thus, need to be fundamentally considered for on-demand and industrial pharmaceutical 3D printing. In the past, it would have seemed like science fiction to design and produce our medicines with 3D printing technology, but it is now certainly becoming reality that will only grow further. This chapter aims to discuss the current regulatory environment for 3D printing mass production and the regulatory gaps and challenges that need to be acknowledged to advance the modern manufacture of medicines.

Keywords 3D printing · Personalized pharmaceuticals · Additive manufacturing · Process validation · Medical devices · Drug delivery systems

1 Introduction

In August 2015, the United States Food and Drug Administration (US FDA) approved the first three-dimensional (3D) printed pharmaceutical drug product [1] and thus, set the benchmark for an industrial revolution of medicine manufacture. Since the completion of the human genome project in the mid 1990s, the concept of personalization is continually being explored in the pharmaceutical arena [2–4]. Traditional carrier systems for drugs (namely tablets, capsules and other dosage

A. Khairuzzaman (✉)
Office of Pharmaceutical Quality, The US FDA, Silver Spring, MD, USA
e-mail: Akm.Khairuzzaman@fda.hhs.gov

A. W. Basit, S. Gaisford (eds.), *3D Printing of Pharmaceuticals*,
AAPS Advances in the Pharmaceutical Sciences Series 31,
https://doi.org/10.1007/978-3-319-90755-0_11

forms) are slowly becoming redundant. Actives such as DNA or other macromolecules are fast developing to become the new generation of treatments. These modes of therapies can be tailored and optimized to cater to individual characteristics, needs and preferences of each individual patient. The rapidly evolving paradigms in medicine, however, is posing a challenge towards the current bulk manufacturing methods of standardized dosage forms. Traditionally, high speed pharmaceutical-grade machineries are used to produce hundreds of thousands of pills, designed with a "one size fits all" approach. These bulk manufacturing techniques must ensure that the drug product is thermally and chemically stable over the period of shelf life (2 years or more). In addition, regulatory standards require each product to be identical with each batch with respect to purity, impurities and standardized quality attributes.

Every year, prospective molecules with promising therapeutic benefits are withdrawn from the development pipeline due to their poor physicochemical instability, despite the extensive formulation research invested to increase product shelf-life. However, these potential candidates could potentially be formulated with an appropriate on-demand manufacturing platform with little or no long-term stability for patient administered immediately after preparation. Although this approach may seem straightforward, and to some extent may be achievable through a simple compounding process, complexities may arise during product design if a specific target release profile and/or precision regarding purity and impurity is desired. More importantly, the regulatory structure of individual countries with respect to manufacturing, distribution and compounding of pharmaceutical products within or outside the hospital and industry should also be fundamentally considered. With such new paradigms, the recent FDA approval of the first 3D printed drug product created enormous attention in the community of personalized medicine. Whilst 3D printing has already demonstrated a commercial success of bulk pharmaceutical product manufacturing, it clearly shows the capability of on-demand manufacturing. In the healthcare arena, although 3D printing has recently introduced itself to the pharmaceutical manufacture of medicines, The Center for Devices and Radiological Health (CDRH) has approved and regulated 3D printed products in the form of medical devices within the last decade [5].

2 The Current Regulatory Environment for 3D Printing Mass Production

2.1 The Code of the US Federal Regulations

Although 3D printing is a relatively new technology, it does not require a unique regulatory pathway. Like any other solid oral dosage forms, a 3D printed drug product must also be manufactured in accordance to the current chemistry, manufacturing and control (CMC) standards as set forth in the 21 CFR 200 s and 300 s and other relevant guidance for 505 b (1) or 505 b (2) types of submissions. It is to be

Table 11.1 US regulatory provisions to submit pharmaceutical drug product applications

505 (b) (1)	New Drugs	The 505(b) (1) regulatory pathway is the traditional New Drug Application (NDA). This pathway is used to obtain the approval of a new drug whose active ingredients have not previously been approved. This pathway requires extensive clinical and non-clinical studies to demonstrate the safety and efficacy of a new molecule for the target indication[a]
505 (b) (2)	New Drug containing similar active ingredient as a previously approved drug.	A 505(b) (2) application is one for which one or more of the investigations relied upon by the applicant for approval "were not conducted by or for the applicant and for which the applicant has not obtained a right of reference or use from the person by or for whom the investigations were conducted" (21 U.S.C.355(b) (2)[b]
505 (j)	Generic drugs (ANDAs)	An abbreviated drug application supported by bioavailability/bioequ ivalence studies comparing the proposed product to the reference listed drug product (RLD)[c]. FD&C 21 U.S.C. 355(j)

[a]21 Code of Federal Regulation (CFR), Part 310
[b]Draft Guidance for Industry: Applications covered by section 505(b) (2)
[c]21 Code of Federal Regulation (CFR), Part 314.3 (b)

noted that the recent FDA approval of the first 3D printed drug product used the existing 505 b (2) regulatory pathway of the Federal Food, Drug, and Cosmetic Act (the Act). In the US, drugs can be approved via one of three regulatory pathways. For new drugs, there are two primary pathways; 505 b (1) and 505 b (2). For the generic drugs, the only available pathway; 505(j), must be used. All these pathways are flexible enough to address new technologies which are summarised in Table 11.1.

All these three provisions provide a fundamental platform for both new and generic drugs irrespective of the manufacturing technology. In addition, there are numerous discipline-specific guidance published from the FDA such as clinical, non-clinical, CMC, statistics, compliance and good manufacturing practices (GMP) which support each of these regulatory platforms.

In the US, medical devices are regulated under the regulatory provision of 510 (k). Such provision can also allow all types of 3D printed medical devices (Class I, II and III). The FDA's CDRH has also recently published a guidance for medical devices entitled "*Technical Considerations for Additive Manufactured Medical Devices Guidance for Industry and Food and Drug Administration Staff*".

2.2 *The International Council for Harmonization*

Since its inception in 1990, the International Council for Harmonization (ICH) has sought to achieve a greater worldwide harmonization on scientific and technical foundations of drug registration. All four ICH topics and their principles, *Quality*, *Safety*, *Efficacy* and *the Multidisciplinary guideline*, can be equally applied to the 3D printing technology for pharmaceutical products. With respect to *Quality*, the ICH supports advancing the regulatory science and innovation in favour of a cleaner, flexible and more efficient manufacturing of pharmaceutical products. This is managed by encouraging industries to adopt new technologies as supported by ICH Q8 (R2), Q9, Q10, and Q11, in conjunction with science and risk based approaches to assure product quality. All other CMC aspects of a 3D printed drug product such as stability, impurities, drug substance and drug product specification, and GMP can still follow the ICH guidelines as outlined in Q1, Q2, Q3, Q6, and Q7.

2.3 The Rest of the World (RoW)

The regulatory environment of pharmaceutical products in the European Union is governed by the European Medicines Agency (EMA). The EMA quality guidelines provide a very detailed instruction for active substances, manufacturing, impurities, specifications (analytical process and validations), excipients, packaging, stability, pharmaceutical development, specific type of products and lifecycle management. For any 3D printed drug products, the principles of all these quality attributes can be similarly applied with additional features that are specific to the 3D printed products. Other regions such as South America and Asia, however, have implemented individual regulations and compendial pharmacopoeias that regulate pharmaceutical products. If used, 3D printing should also be under their regulatory jurisdiction for mass production and distribution within their respective countries.

3 Development Considerations for 3D Printing

3.1 Product Design and Process

As aforementioned in previous chapters, 3D printing is a layer-by-layer process, capable of producing 3D object from digital designs. Unlike traditional pharmaceutical processes (known as subtractive process), 3D printing is a bottom up approach whereby sequential building of thin two-dimension (2D) layers and joining each to the layer below is done over and over by a controlled movement of the print head and/or the base plate [2–4, 6–8]. 3D printing technologies have the flexibility to produce final dosage forms that allow patients to be given personalized regimes such as polypharmacy [3, 4] (see Chap. 1). There are a multitude of 3D printing platforms (see Chap. 2) with technologies utilizing different types of input materials or operating principles. Therefore, the product and process design using this platform greatly depends on the type of 3D printer used and vice versa. Certain product designs are specific to the 3D printer. For example, the first FDA approved 3D printed product, Spritam, is fabricated using a proprietary manufacturing process known as the ZipDose technology and inkjet 3D printing. This method creates porous tablets that dissolve almost instantaneously in the mouth [1]. The physical product design using a traditional pharmaceutical process greatly depends on the dosage form and method of manufacture. In addition, the functional design of a

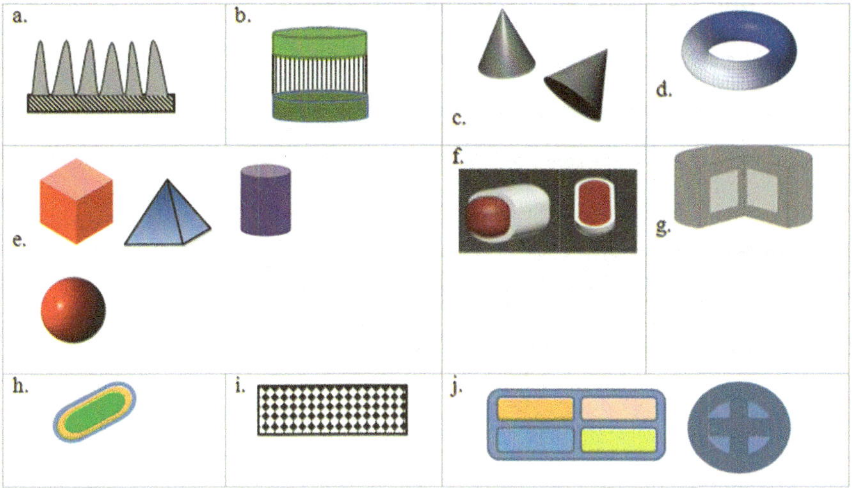

Fig. 11.1 Unmoldable and complex product designs using 3D printing; (**a**) 3D printed miconazole loaded microneedles [9]; (**b**) Acetaminophen tablet built in different gradients [10]; (**c**) Felodipine loaded HME filament [11]; (**d**) Donut hole-shaped pill containing paracetamol, acetaminophen and control release polymers [12]; (**e**) 3D printed drug loaded cube, pyramid, cylinder and sphere [13]; (**f**) Shell core delayed release tablet using FDM 3D printing [14]; (**g**) 3D printed tablet with bound region within the middle of the device [15]; (**h**) a caplet-within-a caplet [16]; (**i**) 3D printed drug delivery patches [17]; and (**j**) a Multi-compartmental tablet [18]

product (e.g. modified release products) greatly depends on the formulation development.

3D printing technology has shown to be amenable in fabricating a variety of established pharmaceutical dosage forms such as tablets, capsules, powder, lozenges, films (2D printing), microneedle transdermal patch and others [2–4]. Using 3D printing product designs, however, can aid the development of complex geometries. Figure 11.1 shows examples of spherical, cylindrical, pyramidal, cubic and multi-layered combinations of 3D printed dosage forms found in the literature, fabricated to explore pharmaceutical feasibility.

Similarly, complex functional design created using a 3D printing technology can affect the drug release of oral dosage forms. If a bimodal drug release mechanism is needed to be built into a formulation (for example, a combination of osmosis controlled and diffusion controlled system for multi drug in a single pill), such design of a product would be difficult to produce using conventional bulk production methods. The deposition of drug substance may require a gradual concentration gradient and other spatial patterns throughout a dosage form to control the rate and amount of drug release from each sub-section. More recent research has demonstrated that 3D printing technologies can produce solid oral dosage form by digitally controlling the arrangement of drugs and excipients precisely in multiple layers. The active pharmaceutical ingredient(s) can be digitally partitioned in multiple layers of various types of functional excipients by varying densities and diffusivities to achieve a

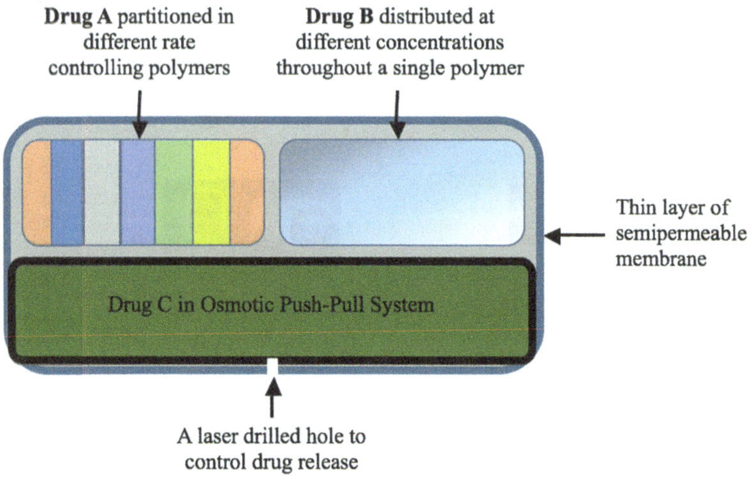

Fig. 11.2 A fixed dose combination of a multicompartmental pill containing three different drugs in one tablet. Drug A is partitioned in various types of polymer vertically which controls the drug release based on dissolution and diffusion controlled mechanism. Drug B is distributed at various concentrations by applying dispersed droplets by the print head throughout the pill with an erosion controlled polymer. Therefore, drug release is achieved following the erosion mechanism. Drug C is an osmotic push-pull system whereby drug release is governed by osmosis

multimodal drug release profile. Figure 11.2 illustrates an example of a complex formulation that is portrayed based on several research publications [19, 20]. These types of complex formulations are unique for site-specific and disease-responsive drug release that has stringent control over *in vivo* biodistribution. These complex designed tablets are considered to be a very thin slab or an individual micro tablet within a tablet, capable of releasing drug either through a pulsatile fashion [21] or through a combination of multiple drug release mechanism. This may include a combination of an osmotic pump, diffusion controlled and/or erosion controlled drug release [22].

They are also unique for drug candidates that require high precision such as narrow therapeutic index drugs, potent drugs, insoluble drugs and drugs that require nanotechnology tailored to the patient [23]. All these examples of complex formulation design can be created and manufactured using 3D printing technology. However, the very fundamental question to ask is whether the product would be administered to a single patient, a group of patients in a clinical or hospital setting, or a population of patients. The answers to these questions are expected to significantly influence the design and process considerations. Firstly, if a 3D printed product is developed for a population of patients and produced using commercial scale manufacturing, all traditional CMC related regulation, therefore, will be applied. Thus, identical final formulation compositions for every unit of the dosage form should be manufactured and will be required to meet the target shelf life and specification. Additional product quality attributes and manufacturing controls may be applied due to the nature of additive manufacturing process.

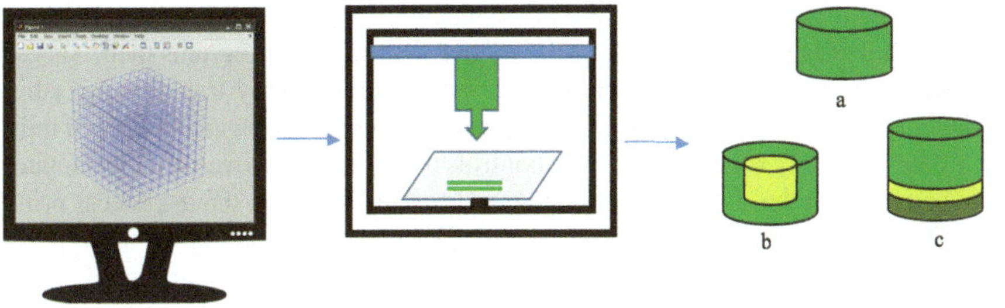

A pharmaceutical tablet is first designed in three-dimensions using a computer aided design (CAD)

The product design file is sent to a 3D printer. The printing program then slices the design in multiple 2D layers. The software controls the print head for a layer-by-layer deposition of the material onto the build plate

Two dimensional layers are fused throughout the process to create a single 3D product such as (a) plain matrix tablet, (b) a tablet within a tablet and (c) a multi-layered tablet containing a single or multiple drugs

Fig. 11.3 Step wise schematic of a typical 3D printing process

Under both circumstances (3D printing for personalized use or mass population), the following factors are generally considered for the development of a 3D printed product:

(a) drug product design matching the target patient vs. target group of patients
(b) specific patient needs (group or individual), e.g. localised drug delivery through 3D printed microneedle transdermal patches
(c) type of the molecule to be delivered and its indication, e.g. biologics may require specific formulation and process design in addition to specific regulatory requirements
(d) target dose and its level of precision
(e) level of personalization
(f) target delivery route
(g) target in vitro drug release and corresponding pharmacokinetic characteristic to be achieved
(h) target quality attributes
(i) target packaging configuration, and
(j) target shelf life

Depending upon the product's internal, external and functional design developed using computer aided design (CAD), a pharmaceutical company may use a variety of 3D printing platforms. Although different types of 3D printers operate on different principles, most 3D printing processes follow the schematic presented in Fig. 11.3 [24–46] and the following stepwise procedures for manufacturing pharmaceutical products:

Step 1: A 3D digital design of the intended product is first created with CAD software.

Step 2: The design is then converted to a machine-readable .stl file format which describes the external surface of a 3D model. The printing programs then "slices" these surfaces into distinct printable 2D layers and digitally transfers layer-by-layer instructions to the printer. The number of layers and thickness of each individual layer may vary. The software controls the print head movement and dictates where to print the support material to provide scaffolding for the in-process print.

Step 3: At this stage, the raw materials are processed. Depending on the type of printing method, raw materials may be processed into granules, dry mixed powder, filaments, or binder solutions to facilitate the printing process. The active drug substance may be incorporated in the granules/powder or in the print fluid. Sometimes a combination of two or more drugs are incorporated.

Step 4: Raw materials are then added and solidified in an automatic, layer-by-layer manner to produce the desired product which is known as the build-up cycle.

Step 5: After printing, products may require additional treatments such as drying, sintering, polishing or other post-processing steps [2]. At this stage, any unprinted materials may be harvested and recycled for continued use in the printing process. If the unprinted harvested material contains any active ingredients, however, it may require further characterisation for possible drug degradation, segregation and other typical physical properties commonly tested for powders.

Once a specific type of printer is selected, the following process considerations are generally given;

(a) Is the 3D printing process suitable for mass production?
(b) Is scale-up a challenge when the technology is transferred from table top research and development (R&D) 3D printers to mass scale production size 3D printers?
(c) Is the manufacturing speed reasonable to meet the market demand?
(d) Is the process capable to reproduce each unit dose consistently and has flexibility?
(e) Does the process have the capability to accommodate a framework for Innovative Pharmaceutical Development, Manufacturing and Quality Assurance (PAT) if necessary?
(f) Is the process operatory friendly and can be compliant with all kinds of qualification and validation in a cGMP environment?
(g) Is the process well controlled by the software? And,
(h) Can the software be validated?

3.2 *Special Considerations for 3D Printing in Pharmaceuticals*

3.2.1 State of Control

A 3D printing manufacturing process should maintain a state of control to provide assurance that the desired product quality is consistently met in a routine commercial set up. Since 3D printing is still new and has less experience in the manufacturing of pharmaceutical products in bulk, certain situations such as sudden software malfunction, uncontrolled change in process variables (e.g. print head voltage gap, print head clog, roller speed, powder bed alignment, print head alignment, laser power, temperature and many others as described later) are needed to be well understood with respect to product homogeneity and quality. Therefore, control for equipment design, product orientation, layer thickness, print head height from the powder bed, printing speed and printing patterns apply to all 3D printing processes.

3D printing manufacturing may not be as speedy as conventional pharmaceutical manufacturing techniques. Appropriate process attributes or ranges need to be identified during development. The ability to detect the process upsets and corrective actions to bring the process back into conformance (e.g. feedback control) will further ensure product quality consistency. To ensure that a process parameter or product attribute cannot move outside the control monitoring system, quality systems should be able to take measurements at a frequency which is appropriate to a specific type of 3D printing used and the length of manufacturing process. Measurement frequency should consider the intrinsic process risk (e. g. higher risk at lower API dose or complex product design as shown in Figs. 11.1 and 11.2), the known process variability, and the required residence time in the equipment to complete transformation (e. g. higher risk at faster throughput or vice versa for extrusion based 3D printing). Therefore, a thorough risk analysis should be employed to ensure that any impact on product quality is understood and appropriately managed in all operating states. Finally, like any other routine commercial pharmaceutical manufacturing, appropriate sampling, testing, quality control procedures and equipment mechanisms to detect and reject materials which are out of specification are also needed.

3.2.2 Raw Material and Intermediate Control

3D printing requires additional raw material control. As 3D printing is a layer-by-layer process whereby each individual layer may be thinly distributed (as low as 10 microns or below), the starting material may undergo significant physical and/or chemical transformations. Thus, the starting material and other inactive ingredients added in the path of the build cycle can have a significant effect on the properties of the finished drug product. For example, in the case of powder bed inkjet 3D printing, if the flow properties and micromeritics of materials are not adequately characterised, variable layer thickness and improper adhesion between layers may occur. Thus, the target product design may be drastically different to the product fabricated at the end of the build cycle. In general, the selection of APIs and excipients greatly depend on their printability and physicochemical stability, in conjunction with the type of 3D printer to be used. For example, material specifications will be different for powder based 3D printing techniques to that of stereolithography [47, 48] (See Chap. 4). Therefore, it is extremely important to maintain a relationship with the raw materials supplier and continuous monitoring of lot to lot variability. This can be achieved by setting appropriate incoming material specifications, the use of applicable validated analytical testing methods and standards (e.g. ISO or ASTM). Typical raw material characteristics may include: micromeritics (e.g. particle size distribution, surface area, shape, surface charge, porosity), flow properties, visco-elastic properties, melting point, glass transition point, thermal conductivity, viscosity (or melt viscosity), impurities, degradation properties, solid state and presence of moisture. For polymeric raw materials, specific characteristics such as monomer mixture/composition, purity, water content, molecular formula, chemical structure, molecular weight distribution, cross linking, molecular weight distribution, glass transition temperatures, melting and crystallization point temperatures, may be applicable. Table 11.2 further lists specific raw material properties for specific types of 3D printing.

Depending on the type of 3D printer used, there could be various types of intermediate materials involved in the path of 3D printing process. For instance, a powder inkjet printing process typically consists of two types of intermediates. Firstly, a powder blend is prepared which may or may not include the API. Secondly, a print fluid is also prepared to bind successive layers together. This solution may or may not contain another active ingredient. For the intermediate blend, rheological properties (e.g. flow), uniformity of the blend as a function of different level of harvesting cycle, moisture and degradation impurities are very critical. For the print fluid intermediate, properties like viscosity, density, composition (single phase vs multi phasic), thermal conductivity, API stability in print fluid and rheology can be critical to characterise. Similarly, for the fused deposition modeling (FDM) printing process, an upstream separate manufacturing process could be involved for the preparation of drug loaded polymeric filament, in which case the filament would be considered as an intermediate. In such case, glass transition temperature of the filament, API uniformity, density, size, thickness and other properties can be important to consider for the downstream extrusion based printing operation.

Raw material and intermediate control approaches should be based on product and process understanding and may include PAT monitoring. The determination of the characteristics of an intermediate product that may or may not be isolated may be more difficult in a 3D printing process due to the limited sampling ports, especially if it is assigned for personalized use. The quality of excipients and intermediates should be linked to the product critical quality attributes and the needs of the process.

3.2.3 Equipment and Software Workflow

It is important to consider equipment control aspects for 3D printing since there are a multitude of 3D printers available whose operating principles are drastically different from each other. The most widely used 3D printers are powder inkjet 3D printers and FDM. Table 11.2 summarises the typical manufacturing risk, raw material and process control for each type of 3D printers.

As described in Fig. 11.3, each 3D printer operates by a software work flow. Without a proper software work flow, the printer is not functional. Therefore, the following software attributes need to be carefully considered during development and in commercial manufacturing.

Table 11.2 Typical manufacturing risk, process control, raw material control and typical quality defects

Type of 3D printer	Typical manufacturing risk	Process control	Raw material control & intermediate material properties	Typical quality defect
Inkjet printing/ binder deposition	Variable layer thickness	Software control, real-time layer thickness monitoring/PAT, base plate speed and powder roll speed, and material feed rate	Particle size and shape, porosity, moisture content and surface charge	Banding[a], leaning[b], and friable tablets
	Print head clog, inconsistent binding, agglomeration between layers, delamination	Real time monitoring of print area density (dpi) through monitoring of voltage gap for each print nozzle, real time monitoring of print fluid fill height, process parameters such as jetting rate and temperature, print head and powder bed speed, and the distance between print head and powder bed	Binder viscosity, binder surface tension, thermal conductivity, specific heat capacity and glass transition	Collapse[c], friable/ delaminated tablets, poor tensile strength, disintegration and dissolution
	Inefficient drying and variability in harvesting (recycled material)	Use of PAT/NIR for monitoring of blend uniformity on harvested material Use of PAT/NIR for monitoring of moisture during drying	Blend uniformity and moisture content of the harvested material, particle size distribution of the harvested blend and flow properties	Content uniformity, weight uniformity, degradation impurities, polymorphic conversion and microbial contamination
	Inconsistent print droplet formation (e.g. drop diameter, drop volume, Reynolds/Weber and Ohnesorge number)	PAT or parametric control (e.g. drop velocity, voltage gap, temperature)	Binder/fluid viscosity, binder surface tension and dynamic viscosity	Friable/delaminated tablets, disintegration and dissolution

(continued)

Table 11.2 (continued)

Type of 3D printer	Typical manufacturing risk	Process control	Raw material control & intermediate material properties	Typical quality defect
Extrusion and fusion/Fused Deposition Modeling (FDM)	Clog, uniformity problem due to undesired melt viscosity. Delamination between layers, improper melting	PAT – use of NIR for real time monitoring of API concentration in layers and entrapped air bubbles, use of laser for individual layer thickness monitoring and monitoring of degradation. Parametric control – Nozzle diameter, pressure, temperature and head speed. Material flow rate, Cooling – rate, temperature Gap between software and hardware, print alignment	Filament uniformity, polymer molecular weight distribution, glass transition temperature, viscoelastic properties, specific heat capacity, thermal conductivity, API melting point (if API is loaded in the extrusion filament) Polymer mechanical properties	Warping[d], Stringing[e], Residuals[f] Split tablet/pill due to delamination Dissolution, Content Uniformity, Degradation Solid state conversion
Stereolithography and selective laser sintering	Variable layer thickness and compositional variation, insistent fusion, laser induced degradation	Penetration depth, critical exposure (energy/area), power of the laser, speed of the laser, beam diameter, powder deposition rate, roller speed relative to powder	Particle size and shape distribution, powder flow properties, polymer miscibility with API, monomer molecular weight, solution concentration, pre- and post-sintering homogeneity, amorphousness	Laser induced degradation, solid state conversion, content uniformity, Warping[g], and Residuals[f]

[a]Banding: ripples on a product's sides caused by vibration in the x–y plane during printing
[b]Leaning: off-axis products caused by drift in the x–y plane during printing
[c]Collapse: loss of porosity caused by sagging layers or excessive mass/energy input
[d]Warping: product distortion caused by thermal expansion or contraction
[e]Stringing: wisps of filament caused by filament elongation during an extruder's off phase
[f]Residuals: un bound powder or uncrosslinked monomer caused by incomplete printing
[g]Warping: product distortion caused by thermal expansion or contraction

(a) Software interaction/alignment with the drug product design: It is likely that during additive manufacturing, the actual product design may shift due to inappropriate software control. Therefore, it is critical to understand the clinical relevant product design parameters and outline a predetermined design parameter range.
(b) Complex drug product design (e.g. Figs. 11.1 and 11.2): A complex design file in a computer can be vulnerable due to errors during file conversion. Therefore, careful considerations need to be given in file format conversions. Once the final design of the product is established, it should be archived and maintained appropriately following the Additive Manufacturing File format (AMF) described in the ISO/ASTM 52915 *Standard specification for additive manufacturing file format (AMF).*

3.2.4 Print Cartridges

Print cartridges may be manufactured with drug loaded formulation in it. However, under the current regulatory paradigm in the US and in the EU, it is not clear if that should be regulated as a drug device combination product or as a pharmaceutical raw material manufacturer. If the ink is a formulated drug (or a formulated binder solution without the presence of drug, or a polymeric base material) that will be used downstream for printing a dosage form, then it is highly likely that it might be considered as an intermediate raw material. In such case, the facility should be considered as a part of the pharmaceutical manufacturing and be subject to standard regulation and GMP. In addition, the content of the print cartridge should comply with standard CMC. On the other hand, the print cartridges are typically part of the 3D equipment. Print cartridges could be designed in different ways and their working principles could be different (e.g. voltage, thermal, ultrasound). Therefore, if the drug loaded (or excipient loaded) print cartridges are supplied at the site of the printing facility, it should be controlled appropriately by the facility where the cartridges will be used to print the final dosage forms.

3.2.5 Process Monitoring: PAT and Possible Real Time Release Testing

3D printing process may utilize real time process monitoring on an industrial scale for better performance and to ensure consistent product quality. Since 3D printing is a continuous additive manufacturing process, on an industrial scale, it is convenient to use process analytical technology with feedback and feed forward controls in place. Because of the nature of the additive manufacturing, each layer of the deposited material contributes to the final quality and product specification. Therefore, if a single layer or some sequential layers are out of its target specification, it would be too late for a corrective action at the end of the build cycle in absence of real time monitoring. PAT can also be implemented on a smaller scale either for on-demand manufacturing for personalized reason, or in drug product pilot scale development for better understanding process design/space linking the product's critical quality attributes. In either case, the implementation of PAT should follow the same principles of the *FDA Guidance for Industry PAT – A Framework for Innovative Pharmaceutical Development, Manufacturing, and Quality Assurance* which is typically applied for pharmaceutical drug product manufacturing using conventional technologies. The use of PAT in the 3D printing process specifically helps in process understanding and establishment of appropriate control strategies. On-line, in-line, or at-line measurement of critical quality attributes provide for control strategies that include real time quality evaluation that is at least equivalent to, or better than, laboratory-based testing on collected samples. In 3D printing, different types of PAT tools can be used. Typically, spectroscopic tools (e.g. NIR, X-ray, IR, Raman, FTIR) are used for real time measurement of API content, uniformity, moisture content, polymorphism, degradation, air entrapment and other properties. Often, temperature sensor, image sensor, ultrasound, hyperspectral imaging and lasers are also used for real-time measurement of melting temperature, individual layer thickness, product geometry, real-time ink level and others. Mathematical models using PAT can also be built for real time measurement of certain quality attributes such as assay, dissolution, impurities and thus for real time batch release without any conventional analytical testing.

3.2.6 Product Collection and Rejection

3D printing is a new platform in pharmaceutical bulk manufacturing. Although this technology was originally introduced to pharmaceuticals for the on-demand manufacturing of personalized medicine, the recent FDA approval of Spritam established a milestone for its use in commercial bulk manufacturing. Therefore, when used, 3D printing process is expected to maintain a state of control like any other conventional pharmaceutical manufacturing. However, there may be situations where products made during the malfunction of either the software or its misalignment with the printer is removed whilst the remainder of the product is retained. Other situations may warrant rejection of the entire batch instead of a portion of a batch. Establishing appropriate criteria for product collection, product rejection, rejection of an entire batch and indicating how or who makes those decisions (either manual or with the help of PAT with inbuilt feed forward logics) helps to ensure the desired quality and consistency of the collected product.

3.2.7 Material Carryover

It is a common occurrence in 3D printing for a certain amount of material to be carried over after the build cycle. Depending on the type of 3D printer used, the amount of material carryover could be significant. For example, in the case of powder inkjet 3D printing process, more than half of powder material could be carried over after the completion of the build cycle. In such a case, the unbound materials are collected following appropriate procedure which could be recycled back to the build cycle operation. Therefore, it is extremely important to characterize these carried over materials for their purity and impurity, including degradation, uniformity, flowability, moisture and other physicochemical properties as appropriate for the 3D printing process.

3.2.8 How to Set Specifications for 3D Printed Drug Products

Setting drug product specifications made by 3D printing should follow the principles of the ICH Q6 guideline. However, it is important to note that this guideline is for routine bulk pharmaceutical product manufacturing only. If 3D printing is used to manufacture any standard conventional dosage forms (e.g. capsules, tablets, parenteral, transdermal patch, ointment, and suspensions) as described under the dosage form nomenclature in compendial pharmacopoeias (such as USP, BP, and JP), a clinically relevant specification accommodating all critical quality attributes related to these dosage form should be established following the guideline. In addition to typical quality attributes in specification such as assay, impurity, content uniformity, and dissolution, additional quality attributes for the developed product made by 3D printing may apply. As for example, if a solid oral tablet is developed using a powder inkjet 3D printing technology with a very porous internal structure that may dictate the fast dissolution of a tablet, then porosity of the tablet could be an additional quality attribute. Such attribute may also impact the tensile strength of the product which is critical to maintain the physical structure of the product from breakage upon shipping and handling. Simulated shipping study of these type of 3D printed dosage form may ensure products integrity upon distribution, shipping and storage.

Setting specification of a conventional drug product (as described above) for personalized use, or for on demand manufacturing in certain situation like warfare, epidemic and drug shortage should also follow the same procedure. However, certain attributes in the specification may vary in a patient by patient and/or situation by situation basis. For example, personalized dosing needs to be set for specific patient or group of patients (e.g. 10 different doses for 10 different patients, or in different combinations for multiple drugs). Similarly, certain quality attributes may not be necessary in situations like warfare where there is an immediate need for a pharmaceutical product in life-saving situations.

3.2.9 Scale Up

If a prototype formulation is developed using a table top inkjet 3D printer which showed promising results in clinical study, the same type of printer, therefore, is needed to be used for large scale manufacturing. In such case, determining scale up factors impacting the product quality could be challenging due to unavailability of prior knowledge of this technology in a commercial setting. Optimal setting and parameters for a single model of machine can vary greatly for different products, as well as between machines of the same model and size even when printing the same product. During the R&D scale pilot development, design of experiments need to be conducted to identify the formulation and process variables that could significantly affect the product quality and performance. The outcomes of such a study could further help the identification of critical process parameters that are directly related to equipment size, type and principle. If a drug product was developed using an extrusion based 3D printer, scaling up to a larger equipment with multiple print heads representing the same table top extrusion based 3D printer will not greatly pose product variations following scale-up. If the nozzles of the industrial scale machine are of a different type to that used in the pilot scale product development study, however, certain parameters such as nozzle diameter, pressure, temperature, head speed, material flow rate, cooling rate, build speed, total energy density and others could be different. This would ultimately affect scale up parameters and would therefore, need to be identified and considered. Similarly, for inkjet 3D printer, if the print heads are of different types, certain printing process parameters such as voltage gap, distance from powder bed, need to be re-established and thus, increase the challenges for scale up to occur. Additional scale up factors with respect to powder handling (e.g. mixing of larger quantity of powder material), different base plate configuration, base plate speed, roller speed, material feed rate, number of recycling of harvested material and post building operations such as different dryer configuration should be measured. Many of these scale-up issues impacting the target quality criteria of each individual layer can be well managed with a real-time monitoring of the established set of physicochemical attributes and its boundaries.

3.2.10 Post-processing and Packaging

3D printing may often involve different types of post-processing unit operations and can ultimately affect the finished pre-cut quality of the product. Different types of 3D printers have different post processing unit operations. For powder bed inkjet 3D printing technology, post-processing unit operations such as drying, harvesting of unused/unbound powder material, recycling and packaging are involved. Other 3D printing operations may also include conventional film coating techniques, tabletting or encapsulation of 3D printed material, drying, curing and packaging. These post-processing operations impacting the final drug product quality may not be directly related to the type of 3D printer used in the upstream operation and can be corrected in with a full understanding of the root cause. Therefore, all possible potential detrimental effects of post-processing should be identified and appropriate mitigation approaches should be taken. All post-processing parameters should be a part of process validations.

3.2.11 Process Validation

Principles of ASTM standards – *ASTM E2537 Validation: Standard Guide for the Application of Continuous Quality Verification to Pharmaceutical and Biopharmaceutical* – can be applied for the process validation of mass production of pharmaceutical products by 3D printing. This guidance describes Continuous

Quality Verification (CQV) as an approach to process validation where a manufacturing process (or supporting utility system) performance is continuously monitored, evaluated and adjusted as necessary. Operation qualification (OQ) of the printing process should include appropriate parameters/characteristics specific to 3D printer used to establish control limit which result in product that meets all the predetermined requirement and specifications. A systematic quality by design approach in the product development stage establishes the relationship between critical 3D printing process parameters and the pre-determined critical quality attributes (CQAs). This contributes to the understanding of the risk and failure modes of the process. The control system can be a helpful reference in real-time quality assurance that CQV will provide. Data from production batches can serve to validate the process and reflect the total system design concept, essentially supporting validation with each manufacturing batch. The FDA Guidance on "General Principles of Software Validation can also be applied to 3D printing software.

3.2.12 Detection and Treatment of Non-conformity

Handling non-conforming materials resulting from deviation is a key component in pharmaceutical quality systems. If a 3D printer is used for mass production at an industrial site, the following aspects are applicable:

There should acceptable procedures for handling deviations, including detection and removal of non-conforming material from the path of 3D printing process. An appropriate definition of the material that are non-confirming must be established. The 3D printing process control/monitoring system should be developed in such a way so that that it can always recognise a normal process and can divert the non-confirming material from the path. Considerations may be based on time or amount of material impacted by deviation (e.g. printer malfunction, material drift, printer alignment, speed) for material rejection. The importance of the raw material specifications, their lot-to-lot variability and the source of variability to the process performance should be considered. Potential failure modes of the detection device (or PAT tool) sampling procedure should also be clearly understood.

3.2.13 Equipment Cleaning

Cleaning validation considerations for 3D printing equipment and systems should be the same as traditional pharmaceutical manufacturing techniques (e.g. tablet compression machine, encapsulator, granulator, blender, fluid bed). For 3D printing manufacturing, either dedicated or non-dedicated equipment may be utilized. Certain parts of 3D printing system, however, could be disposable or for one time use only. Principles for determining the acceptance criteria for cleaning agents, bioburdens, endotoxins and degradation products for the cleaning validation of dedicated equipment are essentially the same as for non-dedicated equipment. The cleaning process should consider the material build, degradation, microbial growth, cleaning agent removal from the 3D printing equipment parts, equipment complexity and product changeover. Additionally, the frequency, length and quantity of cleaning material should also be considered.

4 Regulatory Gaps and Challenges

So far, there is no clear regulatory guideline in any country to differentiate the path for the manufacturing of personalized medicine from bulk manufacturing using 3D printing technology. While the current regulatory and cGMP supports 3D printing manufacturing at an industrial scale with a "one size fits for all" concept, the biggest challenge is implementing the fabrication of personalized pharmaceutical drug products in a hospital, pharmacy, doctor's office or even at home. This requires the consideration of existing regulatory structures in all countries (primarily the West) that control the manufacturing, distribution and compounding of pharmaceutical products at a commercial level. In addition, there has been increasing number of patents on additive manufacturing which also prohibits and/or restricts the use of this technology for such use (See Chap. 10). For example, in last decade, approximately 6800 patent applications were filed on additive manufacturing with the US patent office alone [17]. With this existing paradigm, the additive manufacturing community (both academia and industry) is currently facing the following regulatory gaps for pharmaceutical drug products:

(a) What should be the regulatory structure if we use this technology to manufacture on-demand personalized products: In the US and in a handful of other countries, there are some existing levels of compounding in a hospital setting to fulfil specific patient needs. The use of 3D printing for such a purpose, therefore, could be subject to the jurisdiction of hospital regulation, practice of medicine, and/or the Institutional Review Board (IRB). In the US, if the drugs produced by compounders (outside of hospital) are not registered as outsourcing facilities, they are regulated under section 503B(b)(5) and section 503B(a)(10) and at state level. However, these compounding regulations do not clearly mention the use of 3D printing for compounding in any setting. Therefore, at a minimum, all current compounding regulations should apply for any compounding operation that use 3D printing technology.
(b) Printing at hospitals, pharmacies and clinical study sites: If printing is employed to manufacture products at a clinical site associated with industrial scale product development and clinical studies (e.g. first in human safety study), all existing regulation applies. If printing occurs at the pharmacy or hospital, then concerns are related to (a), as described above. In both cases, the primary concerns are health hazards and safety aspects resulting from the use of raw materials, organic solvents and the possible formation of unknown toxic degradation products. As 3D printing equipment can be portable and could be used to make multiple medical products, other factors may need to be evaluated as part of the

regulatory process. This includes robustness against shipping, changing environmental conditions or the potential for cross-contamination.

(c) Printing at home or at the doctor's office: There is currently no regulatory framework for patients or doctors to print personalized medication at home (or at doctor's office) using 3D printing. In addition to the list of regulatory concerns expressed above, the fundamental gaps are as follows:

   (i) Bulk manufacturing of ink cartridges containing API and excipients, and their distribution: who and how this will be regulated?
   (ii) Who will dispense those cartridges? The doctor or pharmacists?
   (iii) Who will manufacture the 3D printers? Will they be regulated as medical devices?
   (iv) Who will monitor the manufacturing of medications at home?
   (v) What product specification should be applied and how they should be measured?
   (vi) What would the shelf life of the product be?
   (vii) Should the shelf life of the drug loaded print cartridges be monitored?

4.1 The Role of the US FDA in Enhancing 3D Printing in Pharmaceuticals

From the increasing awareness and exploration of 3D printed medical devices, bioprinting and lately, pharmaceutical drug products, the FDA has initiated several internal regulatory science and research projects across the centres (e.g. CDER,[1] CDRH,[2] CBER[3]). This was implemented to advance the understanding of the relationship between material properties and process parameters on product quality for 3D printed drug products, medical device and biologics [5]. By ensuring that the FDA quality experts understand the science and its application to its full capacity, the Agency can encourage the adoption of innovative approaches to pharmaceutical manufacturing, whilst providing meaningful and appropriate regulatory oversight. So far, the FDA has reviewed and regulated 3D printed products under the existing regulations. There has been extensive cross-centre collaboration within the Agency with individual centres forming specific work groups. The CDER has recently established an Emerging Technology Team that work directly with stakeholders to help identify 3D printing specific scientific issues that could hinder progress or uptake. The CDRH has published guidance (mentioned earlier in this chapter) and have previously held public workshops in the year of 2014 to obtain input from the additive manufacturing community. However, with respect to manufacturing personalized medicine using 3D printing, it is extremely difficult to draft a regulatory

[1] Center for Drug Evaluation and Research

[2] Center for Devices and Radiological Health

[3] Center for Biologics Evaluation and Research

guidance that can provide specific prescription for every fabrication and for every individual. At this stage rather, the Agency is more focused to set a common framework that can allow this technology to manufacture 3D printed products under existing regulations.

5 Conclusion

The genomic revolution and the age of personalized medicine has exponentially increased in interest and understanding since its introduction in the last decade. In such a context, 3D printing has already established itself as an innovative manufacturing technique for the fabrication of medical devices. However, 3D printing has shown great promise with respect to the flexibility of producing final dosage forms that allow the administration of personalized regimes to patients. In the past, it would have seemed like science fiction to design and produce our medicines with 3D printing technology, but it is now fast becoming reality – this technology will only grow further.

References

1. United States Food and Drug Administration. Highlights of prescribing information – Spritam. 2015. Available from: http://www.accessdata.fda.gov/drugsatfda_docs/label/2015/207958s000lbl.pdf.
2. Norman J, Madurawe RD, Moore CMV, Khan MA, Khairuzzaman A. A new chapter in pharmaceutical manufacturing: 3D printed drug products. Adv Drug Deliv Rev. 2017;108:39–50.
3. Trenfield SJ, Awad A, Goyanes A, Gaisford S, Basit AW. 3D printing pharmaceuticals: drug development to frontline care. Trends Pharmacol Sci. 2018;39(5):440–5.
4. Awad A, Trenfield SJ, Goyanes A, Gaisford S, Basit AW. Reshaping drug development using 3D printing. Drug Discov Today. 2018; https://doi.org/10.1016/j.drudis.2018.05.025.
5. Prima MD, Coburn J, Hwang D, Kelly J, Khairuzzaman A, Ricles L. Additively manufactured medical products – the FDA perspective. 3D Print Med. 2016;2:1.
6. United States Government Accountability Office. 3D printing: opportunities, challenges, and policy implications of additive manufacturing. 2015. Available from: http://www.gao.gov/assets/680/670960.pdf.
7. Chhaya MP, Poh PS, Balmayor ER, van Griensven M, Schantz JT, Hutmacher DW. Additive manufacturing in biomedical sciences and the need for definitions and norms. Expert Rev Med Devices. 2015;12(5):537–43.
8. International Standards Organization. ISO/ASTM 52900 Additive manufacturing – General principles – Terminology. [under development] Available from: http://www.iso.org/iso/catalogue_detail.htm?csnumber=69669.
9. Prasad LM, Smyth H. 3D printing technologies for drug delivery: a review. Drug Dev Ind Pharm. 2016;42(7):1019–31.
10. Yu DG, Yang XL, Huang WD, Liu J, Wang YG, Xu H. Tablets with material gradients fabricated by three-dimensional printing. J Pharm Sci. 2007;96(9):2446–56.

11. Alhijjaj M, Belton P, Qi S. An investigation into the use of polymer blends to improve the printability of and regulate drug release from pharmaceutical solid dispersion prepared via fused deposition modeling (FDM) 3D printing. Eu J Pharm Biopharm. 2016;108:111–25.
12. Wang J, Goyanes A, Gaisford S, Basit AW. Stereolithographic (SLA) 3D printing of oral modified-release dosage forms. Int J Pharm. 2016;503:207–12.
13. Goyanes A, Martinez PR, Buanz A, Basit A, Gaisford S. Effect of geometry on drug release from 3D printed tablets. Int J Pharm. 2015;494:657–63.
14. Okwuosa TC, Pereira BC, Arafat B, Cieszynska M, Isreb A, Alhnan MA. Fabricating a Shell-Core delayed release tablet using dual FDM 3D printing for patient-centred therapy. Pharm Res. 2017;34:427–37.
15. Acosta-Vélez GF, Wu BM. 3D pharming: direct printing of personalized pharmaceutical tablets. Pol Sci. 2016;1:2.
16. Goyanes A, Wang J, Buanz A, Martinez-Pacheco R, Telford R, Gaisford S, et al. 3D printing of medicines: engineering novel oral devices with unique design and drug release characteristics. Mol Pharm. 2015;12(11):4077–84.
17. Yi HG, Cho YJ, Kang KS, Hong JM, Pati RG, Park MN, et al. A 3D printed local drug delivery patch for pancreatic cancer growth suppression. J Control Release. 2016;238:231–41.
18. Khaled SA, Burley JC, Alexander MR, Yang J, Roberts CJ. 3D printing of tablets containing multiple drugs with defined release profiles. Int J Pharm. 2015;494(2):643–50.
19. Katstra WE, Palazzolo RD, Rowe CW, Giritlioglu B, Teung P, Cima MJ. Oral dosage forms fabricated by three dimensional printing. J Control Release. 2000;66(1):1–9.
20. Sun Y, Soh S. Printing tablets with fully customizable release profiles for personalized medicine. Adv Mater. 2015;27(47):7847–53.
21. Wu BM, Borland SW, Giordano RA, Cima LG, Sachs EM, Cima MJ. Solid free-form fabrication of drug delivery devices. J Control Release. 1996;40(1–2):77–87.
22. Vogel BJ. Intellectual property and additive manufacturing/3D printing: strategies and challenges of applying traditional IP Laws to a transformative technology. Minn J L Sci Tech. 2016;17(2):880–906.
23. Beck RCR, Chaves PS, Goyanes A, Vukosavljevic B, Buanz A, Windbergs A, et al. 3D printed tablets loaded with polymeric nanocapsules: an innovative approach to produce customized drug delivery systems. Int J Pharm. 2017;528(1-2):268–79.
24. Alhnan MA, Okwuosa TC, Sadia M, Wan KW, Ahmed W, Arafat B. Emergence of 3D printed dosage forms: opportunities and challenges. Pharm Res. 2016;33(8):1817–32.
25. Fina F, Goyanes A, Gaisford S, Basit AW. Selective laser sintering (SLS) 3D printing of medicines. Int J Pharm. 2017;529(1-2):285–93.
26. Fina F, Madla CM, Goyanes A, Zhang J, Gaisford S, Basit AW. Fabricating 3D printed orally disintegrating printlets using selective laser sintering. Int J Pharm. 2018;541(1-2):101–7.
27. Fina F, Goyanes A, Madla CM, Awad A, Trenfield SJ, Kuek JM, et al. 3D printing of drug-loaded gyroid lattices using selective laser sintering. Int J Pharm. 2018; https://doi.org/10.1016/j.ijpharm.2018.05.044.
28. Genina N, Boetker JP, Colombo S, Harmankaya N, Rantanen J, Bohr A. Anti-tuberculosis drug combination for controlled oral delivery using 3D printed compartmental dosage forms: from drug product design to in vivo testing. J Control Release. 2017;268:40–8.
29. Gioumouxouzis CI, Katsamenis OL, Bouropoulos N, Fatouros DG. 3D printed oral solid dosage forms containing hydrochlorothiazide for controlled drug delivery. J Drug Deliv Sci Technol. 2017;40:164–71.
30. Goyanes A, Chang H, Sedough D, Hatton GB, Wang J, Buanz A, et al. Fabrication of controlled-release budesonide tablets via desktop (FDM) 3D printing. Int J Pharm. 2015;496(2):414–20.
31. Goyanes A, Det-Amornrat U, Wang J, Basit AW, Gaisford S. 3D scanning and 3D printing as innovative technologies for fabricating personalized topical drug delivery systems. J Control Release. 2016;234:41–8.

32. Goyanes A, Kobayashi M, Martínez-Pacheco R, Gaisford S, Basit AW. Fused-filament 3D printing of drug products: microstructure analysis and drug release characteristics of PVA-based caplets. Int J Pharm. 2016;514(1):290–5.
33. Goyanes A, Fina F, Martorana A, Sedough D, Gaisford S, Basit AW. Development of modified release 3D printed tablets (printlets) with pharmaceutical excipients using additive manufacturing. Int J Pharm. 2017;527(1-2):21–30.
34. Goyanes A, Scarpa M, Kamlow M, Gaisford S, Basit AW, Orlu M. Patient acceptability of 3D printed medicines. Int J Pharm. 2017;530(1-2):71–8.
35. Goyanes A, Fernández-Ferreiro A, Majeed A, Gomez-Lado N, Awad A, Luaces-Rodríguez A, et al. PET/CT imaging of 3D printed devices in the gastrointestinal tract of rodents. Int J Pharm. 2018;536(1):158–64.
36. Awad A, Trenfield SJ, Gaisford S, Basit AW. 3D printed medicines: A new branch of digital healthcare. Int J Pharm. 2018;548(1):586–96.
37. Goyanes A, Buanz AB, Hatton GB, Gaisford S, Basit AW. 3D printing of modified-release aminosalicylate (4-ASA and 5-ASA) tablets. Eur J Pharm Biopharm. 2015;89:157–62.
38. Melocchi A, Parietti F, Maroni A, Foppoli A, Gazzaniga A, Zema L. Hot-melt extruded filaments based on pharmaceutical grade polymers for 3D printing by fused deposition modeling. Int J Pharm. 2016;509(1-2):255–63.
39. Okwuosa TC, Stefaniak S, Arafat B, Isreb A, Wan KW, Alhnan MA. A lower temperature FDM 3D printing for the manufacture of patient-specific immediate release tablets. Pharm Res. 2016;33(11):2704–12.
40. Muwaffak Z, Goyanes A, Clark V, Basit AW, Hilton ST, Gaisford S. Patient-specific 3D scanned and 3D printed antimicrobial polycaprolactone wound dressings. Int J Pharm. 2017;527(1–2):161–70.
41. Sadia M, Sośnicka A, Arafat B, Isreb A, Ahmed W, Kelarakis A, et al. Adaptation of pharmaceutical excipients to FDM 3D printing for the fabrication of patient-tailored immediate release tablets. Int J Pharm. 2016;513(1-2):659–68.
42. Skowyra J, Pietrzak K, Alhnan MA. Fabrication of extended-release patient-tailored prednisolone tablets via fused deposition modelling (FDM) 3D printing. Eur J Pharm Sci. 2015;68:11–7.
43. Solanki NG, Tahsin M, Shah AV, Serajuddin ATM. Formulation of 3D printed tablet for rapid drug release by fused deposition modeling: screening polymers for drug release, drug-polymer miscibility and printability. J Pharm Sci. 2018;107(1):390–401.
44. Goyanes A, Buanz AB, Basit AW, Gaisford S. Fused-filament 3D printing (3DP) for fabrication of tablets. Int J Pharm. 2014;476(1-2):88–92.
45. Kollamaram G, Croker DM, Walker GM, Goyanes A, Basit AW, Gaisford S. Low temperature fused deposition modeling (FDM) 3D printing of thermolabile drugs. Int J Pharm. 2018;545(1-2):144–52.
46. Zhang J, Yang W, Vo AW, Feng X, Ye X, Kim DW, Repka MA. Hydroxypropyl methylcellulose-based controlled release dosage by melt extrusion and 3D printing: structure and drug release correlation. Carbohydr Polym 2017;177:49-57.
47. Martinez PR, Goyanes A, Basit AW, Gaisford S. Fabrication of drug-loaded hydrogels with stereolithographic 3D printing. Int J Pharm. 2017;532:313–7. https://doi.org/10.1016/j.ijpharm.2017.09.003.
48. Martinez PR, Goyanes A, Basit AW, Gaisford S. Influence of geometry on the drug release profiles of stereolithographic (SLA) 3D printed tablets. AAPS PharmSciTech. 2018; https://doi.org/10.1208/s12249-018-1075-3.

Index

A. W. Basit, S. Gaisford (eds.), *3D Printing of Pharmaceuticals*, AAPS Advances in the Pharmaceutical Sciences Series 31, https://doi.org/10.1007/978-3-319-90755-0

The manufacturer's authorised representative in the EU is Springer Nature Customer Service Centre GmbH, Europaplatz 3, 69115 Heidelberg, Germany. If you have any concerns regarding our products, please contact ProductSafety@springernature.com

Printed and bound by CPI Group (UK) Ltd, Croydon, CR0 4YY
30/09/2026
02225230-0003